Fundamental Theories of Physics

Volume 182

The international monograph series “Fundamental Theories of Physics” aims to stretch the boundaries of mainstream physics by clarifying and developing the theoretical and conceptual framework of physics and by applying it to a wide range of interdisciplinary scientific fields. Original contributions in well-established fields such as Quantum Physics, Relativity Theory, Cosmology, Quantum Field Theory, Statistical Mechanics and Nonlinear Dynamics are welcome. The series also provides a forum for non-conventional approaches to these fields. Publications should present new and promising ideas, with prospects for their further development, and carefully show how they connect to conventional views of the topic. Although the aim of this series is to go beyond established mainstream physics, a high profile and open-minded Editorial Board will evaluate all contributions carefully to ensure a high scientific standard.

More information about this series at http://www.springer.com/series/6001

Maurice A. de Gosson

Born–Jordan Quantization

Theory and Applications

Maurice A. de Gosson
Universität Wien
Vienna
Austria

ISSN 0168-1222 ISSN 2365-6425 (electronic)
Fundamental Theories of Physics
ISBN 978-3-319-27900-8 ISBN 978-3-319-27902-2 (eBook)
DOI 10.1007/978-3-319-27902-2

Library of Congress Control Number: 2015958863

Printed on acid-free paper

This Springer imprint is published by SpringerNature
The registered company is Springer International Publishing AG Switzerland

To My Lady Marmotte

Preface

What is Born–Jordan quantization? And what is it supposed to be good for? Well, it might very well be that Born–Jordan quantization and the associated operator calculus provide us with the only physically correct quantization scheme, as opposed to the Weyl quantization commonly used in physics. Already this fact is sufficient to motivate a comprehensive mathematical study of the topic. Another motivation is that very recent and ongoing work shows that the Born and Jordan approach provides better results in the study of spectrograms in time–frequency and signal analysis, by damping unwanted interference effects. To understand what Born–Jordan operators are about one has to go back to the early years of quantum mechanics, where a rule for quantizing monomials was proposed by Max Born and Pascual Jordan in 1925 following Werner Heisenberg's paper which inaugurated what is nowadays called matrix mechanics; in this paper Heisenberg proposed the idea of a quantum theoretical reinterpretation of the notion of classical observable. One year later, Erwin Schrödinger proposed his eponymous equation describing the time evolution of de Broglie's wavefunctions, and showed that his approach led to the same predictions as Heisenberg's matrix mechanics, thus proving the uniqueness of quantum mechanics as a new theory. The quantum rule proposed by Born and Jordan was quickly superseded, mainly for mathematical reasons, by another rule due to Hermann Weyl, which became de facto the preferred quantization in physics, and thus leading to two different quantum mechanics: Heisenberg's matrix on the one side, and Schrödinger's wavemechanics on the other side.

For all these reasons we believe that this new pseudo-differential calculus deserves to be studied; the present work is an introduction to the topic, which is still in its infancy. We do hope that it will trigger interest among researchers and students. One more word: this book was primarily written for quantum physicists and mathematicians interested in quantum mechanics. However, it might also be of interest to specialists working in signal theory and time–frequency analysis: it suffices to replace everywhere $\hbar$ with $1/2\pi$ and x with t.

I wrote the draft of this book in one week (sometimes during the spring 2015), it was actually a skeleton, and a rickety one, but it contained, with a few exceptions, the main lines of the final version. But fine tuning and putting flesh on the bones was a harder task. So it is both my duty and great pleasure to thank the following mathematicians and physicists for encouragement and stimulating conversations: Paolo Boggiatto (Turin), Elena Cordero (Turin), Glen Dennis (London), Hans Feichtinger (Vienna), Serge de Gosson (Stockholm), Basil Hiley (London), Franz Luef (Trondheim), Fabio Nicola (Turin), Luigi Rodino (Turin), Michael Ruzhansky (London) and Ville Turunen (Esbo).

Special thanks to my beloved wife, Charlyne de Gosson, who has rearranged and corrected the Bibliography, and to my colleague and friend, Glen Dennis, for having read the manuscript with great attention and pointing out numerous typos and errors.

This work has been funded by the grant P27773-N23 of the Austrian *Fonds zur Förderung der wissenschaftlichen Forschung* (FWF).

Vienna, Argelès-sur-Mer, Stockholm Maurice A. de Gosson
Summer 2015

Contents

Chapter 1
Introduction

1.1 From Quantum Theory to Quantum Mechanics

December 14, 1900, is usually regarded as the official date of birth of quantum theory, because on that day Max Planck presented a memoir at a meeting of the Physical Society of Berlin in which he solved the enigma of the blackbody spectrum by introducing a new, fundamental, constant of Nature[1]:

> We therefore regard—and this is the most essential point of the entire calculation—energy to be composed of a very definite number of equal finite packages, making use for that purpose of a natural constant $h = 6.5 \times 10^{-27}$ ergsec.

Planck was a Scientific of the old German school and had certainly not wanted to challenge in such a drastic way the achievements of classical physics, which was believed at that time to have attained a level of almost perfection. There was still, however, one little unsolved problem that had already puzzled Planck's teacher, Gustav Kirchhoff. That problem concerned heat radiation inside a "black-body", which is an ideal absorber of radiation, for instance a hollow sphere with a tiny hole. Any radiation which happens to go in through the hole will be trapped inside and bounce around until it is finally absorbed. Kirchhoff had postulated in 1860 the existence of a function depending only on temperature and frequency and describing the emission of radiation of a heated black-body, but he hadn't been able to find any explicit form for such a function. Ultimately, at the very end of 19th century, one was left with two formulas: Wien's formula which was rather accurate on the high-frequency side of the emission curve, and the so-called Rayleigh–Jeans law, which agreed well with the predictions of classical thermodynamics in the infrared range. Unfortunately both laws were totally incompatible outside their respective domains of validity; all attempts to make them match had lamentably failed; even worse Rayleigh–Jean's law lead to catastrophic divergences for high frequencies (the "ultraviolet catastrophe"). Planck, who was the successor of Kirchhoff at Berlin

[1] For much more on of the historical development of quantum theory and quantum mechanics see the authoritative book by Jammer [4].

M.A. de Gosson, *Born–Jordan Quantization*, Fundamental Theories of Physics 182, DOI 10.1007/978-3-319-27902-2_1

University, has started working on the puzzle from 1895, but the breakthrough did not come before October 7, 1900, when Planck realized that he had to find a formula which for large frequencies and small temperatures agrees with Wien's formula, but for small frequencies and large temperatures reduces to a proportionality of the energy density with the temperature. He postulated the interpolation formula

$$(u/\nu, T) = \frac{A\nu^3}{\exp(B\nu/T) - 1}$$

(later to be called "Planck's law of radiation") where A and B were constants to be determined experimentally, and presented it at a meeting of the physical Society on October 19, 1900. Planck's formula seemed to be correct; it was checked against experimental results which reported total agreement. This was a great success: Planck's formula agreed beautifully with the measurements of black-body radiation. In 1918, Planck was awarded the Nobel Prize in physics for this discovery. His discovery was the starting point of a new era in physics, which has led to tremendous applications in practically every area of contemporary Science; it has also deeply influenced the evolution of mathematics, particularly functional analysis and geometry. In fact, Planck initiated what is today called "quantum theory"; it took however another quarter of century before this theory got firm mathematical foundations and earned the name of "quantum mechanics". The first rigorous exposition of quantum mechanics was written by Max Born and Pascual Jordan [1] in 1925; elaborating on a paper by Werner Heisenberg, they proposed a "quantization rule" for classical observables; this rule made possible the calculation of atomic spectra. In the simple case of a monomial in the position and momentum variables x and p Born and Jordan's rule read

$$x^r p^s \longrightarrow \frac{1}{s+1} \sum_{k=0}^{s} \widehat{p}^{s-k} \widehat{x}^r \widehat{p}^k$$

where $\widehat{x}$ and $\widehat{p}$ are *operators* satisfying the commutation rule $\widehat{x}\widehat{p} - \widehat{p}\widehat{x} = ih/2\pi$ where h is Planck's constant. This paper was followed by another paper [2] with Heisenberg himself as a coauthor; this second foundational work developed a Hamiltonian mechanics of the atom in a completely new quantum non-commutative format, the "matrix mechanics", still in use today. In the meantime, Louis de Broglie had postulated in his 1924 Ph.D. thesis the wave nature of electrons and suggested that *all* matter has wave properties; in fact, he proposed that to a particle with mass m and velocity v should be associated a wave with wavelength

$$\lambda = \frac{h}{mv}.$$

De Broglie was awarded the Nobel Prize for Physics in 1929 after experimental results had shown his insight was right, and clearly demonstrated the wave-like behavior of matter. De Broglie's theory set the basis of "wave mechanics"; it is at this stage that Erwin Schrödinger enters the scene. In the autumn of 1925 he was invited by Peter

Debye to give a talk at a seminar in Zürich on de Broglie's theory. Intrigued by a question of Debye, who asked

> ...but if matter has wave properties, what is the equation governing these waves?

Schrödinger set out to find a "wave equation". He left for holiday in a mountain cabin in the Alps just before Christmas 1925, and when he returned on 9 January, elaborating on the Hamilton–Jacobi theory, he had postulated the equation

$$\frac{\partial^2\psi}{\partial x^2} + \frac{8\pi^2 m}{h^2}(E - V)\psi = 0$$

satisfied by a stationary matter wave; he later generalized his result to the time-dependent case, which led him to the famous equation

$$i\hbar\frac{\partial\psi}{\partial t} = -\frac{\hbar^2}{2m}\frac{\partial^2\psi}{\partial x^2} + V(x)\psi$$

governing the evolution of the de Broglie waves. The paper [5] in *Annalen der Physik in* which Schrödinger published his results set the foundations of an apparently alternative quantum mechanics, fundamentally different of that of Born, Heisenberg, and Jordan. It was however soon proven that matrix mechanics and wavemechanics were actually equivalent theories: they made the same physical predictions, and one could pass from the one to the other using a simple procedure (see Casado [3] for an account of this equivalence, which was first rigorously proved by John von Neumann). So far, so good: Born, Heisenberg, Jordan, and Schrödinger had created, using different approaches and methods, a new theory, *quantum mechanics*. This theory is two-faced: one can work in either the Heisenberg picture (matrices) or in the Schrödinger picture (waves). The success of this theory was, and still is, undisputed. The story of the foundations of quantum mechanics did however not end with these two theories. Hermann Weyl, who was at that time professor of mathematics at the ETH Zürich, and deeply influenced by the Göttingen school, published in 1926 a paper [6]. In this paper, soon followed by a monograph [7], Weyl used group-theoretical methods to formalize quantum mechanics. He had kept up a correspondence with Born and Jordan, which led him to try to combine the Heisenberg and Schrödinger pictures of quantum mechanics, using abstract methods. Still, Weyl was aware of the fact that wavemechanics was more suitable for physicists; this led him to write in the introduction of his monograph

> ...[Schrödinger's approach] seems to me less cogent, but it leads more quickly to the fundamental principles of quantum mechanics...

Weyl's ideas were met with mixed feelings; while Heisenberg was quite enthusiastic in his recension of Weyl's monograph, Paul Ehrenfest spoke about *Gruppenpest* to emphasize his opposition to Weyl's group-theoretical arguments (but still showed some interest in the new ideas). Without rejecting Weyl's ideas Schrödinger was sceptic and advised him to clarify the physical foundations of quantum mecahnics.

Now, what is of particular interest to us, is that Weyl proposed the following very general quantization rule: to a classical observable $a(x, p)$ depending on the position an momentum coordinates, one should associate an operator defined by

$$a_{\mathrm{Weyl}}(\widehat{x}, \widehat{p}) = \frac{1}{2\pi\hbar} \int e^{\frac{i}{\hbar}(x\widehat{x}+p\widehat{p})} Fa(x, p) dp dx$$

where F is the Fourier transform; this operator is formally obtained by replacing the variables x and p in the Fourier inversion formula by the non-commuting variables $\widehat{x}$ and $\widehat{p}$ used by Born, Jordan, and Heisenberg. Some easy algebra shows that if we quantize a classical Hamiltonian function

$$H = \frac{p^2}{2m} + V(x)$$

using Weyl's rule one obtains the operator

$$H(\widehat{x}, \widehat{p}) = -\frac{\hbar^2}{2m} \frac{\partial^2}{\partial x^2} + V(x)$$

which is the same as the one appearing in Schrödinger's equation. So far, so good. The rub comes from the following observation: if we apply Weyl's rule to the monomials $x^r p^s$ considered by Born, Jordan, and Heisenberg, we get the correspondence

$$x^r p^s \longrightarrow \frac{1}{2^s} \sum_{k=0}^{s} \binom{s}{k} \widehat{p}^{s-k} \widehat{x}^r \widehat{p}^k$$

where the $\binom{s}{k}$ are the binomial coefficients; as is immediately seen by simple inspection, this rule is fundamentally different from Born and Jordan's quantization rule, as soon as $r, s \geq 2$. Thus, if one wants to extend the Schrödinger picture to observables which are *arbitrary* functions of the variables x and p one obtains two different results depending on which quantization rule one uses. This fact has the following unwanted consequence: if one uses Weyl quantization, the Born–Jordan and Schrödinger pictures are *no longer equivalent*, and we thus have two different quantum mechanics. It turns out that Weyl quantization has become the preferred, not to say the only, quantization procedure in modern quantum mechanics. This has many historical reasons we will not try to retrace; one of them is that the Weyl quantization rule is easy to implement mathematically, while one does not see immediately how to apply the Born–Jordan rule to arbitrary observables; as we will see in this book the general formula for the Born–Jordan quantization is

$$a_{\mathrm{BJ}}(\widehat{x}, \widehat{p}) = \frac{1}{2\pi\hbar} \int e^{\frac{i}{\hbar}(x\widehat{x}+p\widehat{p})} Fa(x, p) \frac{\sin(px/2\hbar)}{px/2\hbar} dp dx$$

and this formula is far from being "obvious" ; in particular one does not see where the term $\sin(px/2\pi\hbar)/(px/2\pi\hbar)$ comes from.

1.2 What We Do in This Book

This book consists of three parts, which can be to a great extent be read independently. The logical structure is linear: we have avoided as much as possible to refer, in a given chapter, to results that will be proven in one of the subsequent chapters. This being said, we have made every effort to make each chapter as self-contained as possible.

Here is a short and concise description of the contents:

- Part 1 (Chaps. 2, 3, 4 and 5) is an introduction to Born–Jordan quantization from the physicist's point of view. In Chap. 2 we discuss the problem of quantization in general, and mention the main conceptual difficulties. In Chap. 3 we review the theory of quantization of monomials, which is still a subject a current research, and contains many technical subtleties. In Chap. 4, having the Schrödinger equation in mind, we review the basic ideas of classical mechanics with an emphasis on the symplectic formulation of the Hamiltonian approach. Finally, in Chap. 5, we show that Born–Jordan quantization and the Schrödinger equation for arbitrary observables can be motivated if one makes a simple physical assumption on the short-time wavefunction. This will lead us to a quantization formula which is at the basis of the mathematical theory of Born–Jordan quantization from the pseudo-differential point of view.
- In Part 2 (Chaps. 6, 7, 8, 9, 10 and 11) we develop the mathematics of Born–Jordan quantization. In Chap. 6, we review Weyl quantization, from the point of view of harmonic analysis. In addition to its intrinsic interest Weyl quantization is the easiest way to access Born–Jordan quantization. In Chap. 7 we introduce the "Cohen class" and related objects as the Wigner and ambiguity functions. The results developed in this chapter are essential for the transition from Weyl quantization to the more complicated Born–Jordan quantization. Chapter 8 is central; we give there a rigorous definition of the Born–Jordan quantization procedure, by selecting a particular element of the Cohen class. The operator calculus thus defined is studied. Chapter 9 is technical; there we review the theory of Shubin's parameter-dependent pseudo-differential operators, which will allow us in Chap. 10 to develop the theory of Born–Jordan operators from the pseudo-differential viewpoint. Finally, in Chap. 11, we study the so important notion of *weak value* from the point of view of Born–Jordan quantization, and discuss the so-called "reconstruction problem".
- Part 3 is devoted to some topics which are mathematically more "advanced". In Chaps. 12 and 13 we introduce the theory of the metaplectic group and of its variants; we apply this theory to the study of symplectic covariance both in the Weyl and the Born–Jordan case. Finally, in Chap. 14 we study boundedness properties of Born–Jordan operators in some functional spaces (including Feichtinger's modulation spaces); this requires the introduction of new global symbol classes. This last chapter is concurrent with ongoing research in functional analysis and time-frequency analysis; it thus has a sketchy form.

References

1. M. Born, P. Jordan, Zur Quantenmechanik. Zeits. Physik **34**, 858–888 (1925)
2. M. Born, W. Heisenberg, P. Jordan, Zur Quantenmechanik II. Z. Physik **35**, 557–615 (1925); English translation in: M. Jammer, *The Conceptual Development of Quantum Mechanics*, (McGraw-Hill, New York, 1966); 2nd edn. (American Institute of Physics, New York 1989)
3. C.M.M. Casado, A brief history of the mathematical equivalence between the two quantum mechanics. Lat. Am. J. Phys. Educ. **2**(2), 152–155 (2008)
4. M. Jammer, *The Conceptual Development of Quantum Mechanics*. (McGraw-Hill, New York, 1966)
5. E. Schrödinger, Quantisierung als Eigenwertproblem. Annalen der Physik **384**(4), 273–376 (1926)
6. H. Weyl, Quantenmechanik und Gruppentheorie. Zeitschrift für Physik **46** (1927)
7. H. Weyl, *The Theory of Groups and Quantum Mechanics*, translated from the 2nd German edition by H.P. Robertson, (Dutten, New York, 1931)

Part I
Born–Jordan Quantization: Physical Motivation

Chapter 2
On the Quantization Problem

2.1 Introduction

In 1925 Max Born and Pascual Jordan set out to give a rigorous mathematical basis to Werner Heisenberg's newly born "matrix mechanics". This led them led to state a quantization rule for monomials; that rule associates to the product $x^r p^s$ the operator

$$\mathrm{Op}_{\mathrm{BJ}}(x^r p^s) = \frac{1}{s+1} \sum_{k=0}^{s} \widehat{p}^{s-k} \widehat{x}^r \widehat{p}^k \tag{2.1}$$

where $\widehat{x}$ and $\widehat{p}$ are operators satisfying the canonical commutation relation $[\widehat{x}, \widehat{p}] = i\hbar$. For historical and technical reasons we do not discuss here, Born and Jordan's rule was quickly superseded by a more general rule proposed by Hermann Weyl. Elaborating on the Fourier inversion formula

$$a(x, p) = \frac{1}{2\pi\hbar} \int e^{\frac{i}{\hbar}(x_0 x + p_0 p)} Fa(x_0, p_0) dp_0 dx_0$$

Weyl defined the operator $\mathrm{Op}_{\mathrm{W}}(a)$ associated to an observable (or "symbol") a by formally replacing x and p by $\widehat{x}$ and $\widehat{p}$ in the formula above:

$$\mathrm{Op}_{\mathrm{W}}(a) = \frac{1}{2\pi\hbar} \int e^{\frac{i}{\hbar}(x_0 \widehat{x} + p_0 \widehat{p})} Fa(x_0, p_0) dp_0 dx_0. \tag{2.2}$$

McCoy showed in [15] Weyl's rule leads to the formula

$$\mathrm{Op}_{\mathrm{W}}(x^r p^s) = \frac{1}{2^s} \sum_{k=0}^{s} \binom{s}{k} \widehat{p}^{s-k} \widehat{x}^r \widehat{p}^k \tag{2.3}$$

which is different from Born and Jordan's rule as soon as $r, s \geq 2$ (they however coincide when $r = s = 1$, leading in both cases to the operator $\frac{1}{2}(\widehat{x}\widehat{p} + \widehat{p}\widehat{x})$).

M.A. de Gosson, *Born–Jordan Quantization*, Fundamental Theories
of Physics 182, DOI 10.1007/978-3-319-27902-2_2

The Weyl rule was rediscovered and developed in the 1970s by mathematicians working on the theory of pseudo-differential operators and partial differential equations. It turns out that the Weyl quantization rule is mathematically speaking: is very attractive because of its simplicity; in addition it enjoys a very interesting symmetry property (symplectic covariance; i.e. covariance under linear canonical transformations). It is also intimately related to the Wigner transform, which allows a phase space representation of quantum mechanics. The resulting "Weyl–Wigner" formalism is a well-studied topic in both mathematics and quantum mechanics. So far so good. However, an inconsistency arises when Weyl quantization is used. It comes from the following fact: it is conventional wisdom in physics that the Schrödinger and Heisenberg pictures of quantum mechanics are equivalent (the Schrödinger picture is based on Schrödinger's equation which predicts the time-evolution of the quantum state, and the Heisenberg picture views states as constant in time, and considers the observable to evolve). But, and this has been unnoticed, for this equivalence to hold we *must* use the Born–Jordan scheme, and this because the Heisenberg picture breaks down if we use any other quantization rule. That is, the Schrödinger and Heisenberg pictures are inequivalent if one uses Weyl quantization (or any other ordering rule for that). We are thus left with only one possible conclusion, which might be unwelcome for many physicists: *the right quantization rule for observables is that proposed in 1925 by Born and Jordan.*

From a mathematical point of view, the Born–Jordan pseudo-differential operators are obtained as follows. There are infinitely many ways to associate an operator to a given symbol (or "classical observable") a. For instance, one can use the Kohn–Nirenberg prescription

$$A_{\mathrm{KN}}\psi(x) = \left(\tfrac{1}{2\pi\hbar}\right)^n \iint e^{\frac{i}{\hbar}p(x-y)}a(x,p)\psi(y)d^npd^ny$$

which is very popular among mathematicians working in the theory of partial differential equations, or in time-frequency analysis. Or one can use the Weyl prescription, which is given in pseudo-differential form by the formula

$$A_{\mathrm{W}}\psi(x) = \left(\tfrac{1}{2\pi\hbar}\right)^n \iint e^{\frac{i}{\hbar}(x-y)p}a(\tfrac{1}{2}(x+y),p)\psi(y)d^npd^ny;$$

the latter is very popular among physicists for the reasons discussed above. There is also the anti-normal ordering, which we just mention in passing (it is not widely used). And then, there is the so-called Shubin prescription: for every real number τ one associates a pseudo-differential operator A_τ to the symbol a by the formula

$$A_\tau\psi(x) = \left(\tfrac{1}{2\pi\hbar}\right)^n \iint e^{\frac{i}{\hbar}(x-y)p}a(((1-\tau)x+\tau y),p)\psi(y)d^npd^ny.$$

Obviously, choosing $\tau = 1$ one recovers the Kohn–Nirenberg operator A_{KN}, and choosing $\tau = \frac{1}{2}$ one recovers the Weyl operator A_{W}, so the Shubin operators are just a generalization of known schemes. Now, we *define* the Born–Jordan operator A_{BJ} with symbol a as the average

$$A_{\mathrm{BJ}} = \int_0^1 A_\tau d\tau$$

of all Shubin operators A_τ on the interval [0, 1]; this formula should be interpreted as

$$A_{\mathrm{BJ}}\psi(x) = \int_0^1 A_\tau \psi(x) d\tau$$

for $\psi \in \mathcal{S}(\mathbb{R}^n)$. This definition leads to a completely new pseudo-differential calculus, whose properties are different from those of the operators A_τ (and hence, in particular, from those of the Weyl operator $\widehat{A}_{\mathrm{W}}$). For instance, as opposed to what happens with Weyl or Shubin calculus, it is not obvious that every continuous operator $\mathcal{S}(\mathbb{R}^n) \longrightarrow \mathcal{S}'(\mathbb{R}^n)$ can be represented as a Born–Jordan operator A_{BJ}; the usual argument using Schwartz kernel theorem does not work here (put differently "there might be quantum observables which have no classical analogue"). It also turns out that in Born–Jordan quantization the zero operator can correspond to a non-zero symbol; this particularity raises concerns about the uniqueness of "dequantization"; these matters will be studied in detail.

2.2 The Ordering Problem

Already in the early days of quantum mechanics physicists were confronted with the ordering problem for products of observables (i.e. of symbols, in mathematical language). While it was agreed that the correspondence rule

$$x_j \longrightarrow x_j, \quad p_j \longrightarrow -i\hbar\partial/\partial x_j$$

could be successfully be applied to the position and momentum variables, thus turning the Hamiltonian function

$$H = \sum_{j=1}^{n} \frac{1}{2m_j} p_j^2 + V(x_1, .., x_n) \tag{2.4}$$

into the partial differential operator

$$\widehat{H} = \sum_{j=1}^{n} -\frac{\hbar^2}{2m_j}\frac{\partial^2}{\partial x_j^2} + V(x_1, .., x_n) \tag{2.5}$$

it quickly became apparent that these rules lead to ambiguities when applied to more general observables involving products of the variables x_j and p_j. For instance, what should the operator corresponding to the magnetic Hamiltonian

$$H = \sum_{j=1}^{n} \frac{1}{2m_j} \left(p_j - A_j(x_1, .., x_n)\right)^2 + V(x_1, .., x_n) \tag{2.6}$$

be? Even in the simple case of the product $x_j p_j = p_j x_j$ the correspondence rule led to the a priori equally good answers $-i\hbar x_j \partial/\partial x_j$ and $-i\hbar(\partial/\partial x_j)x_j$ which differ by the quantity $i\hbar$; things became even more complicated when one came (empirically) to the conclusion that the right answer should in fact be the "average rule"

$$x_j p_j \longrightarrow -\tfrac{1}{2} i\hbar \left(x_j \frac{\partial}{\partial x_j} + \frac{\partial}{\partial x_j} x_j \right) \tag{2.7}$$

corresponding to the splitting $x_j p_j = \frac{1}{2}(x_j p_j + p_j x_j)$. To better understand the issue, we have to go back a few years in time, to 1925. That year Heisenberg wrote a seminal paper [13] which defined what we today call "matrix mechanics"; in an attempt to understand Heisenberg's ideas, and to put them on a firm mathematical basis, Born and Jordan [1] wrote a comprehensive paper where they addressed the ordering problem: assume that some quantization process associated to the canonical variables x (position) and p (momentum) two operators $\widehat{x}$ and $\widehat{p}$ satisfying Max Born's canonical commutation rule $\widehat{x}\widehat{p} - \widehat{p}\widehat{x} = i\hbar$. A natural and simple choice (but of course not the only possible one) is to choose the unbounded operators on $\mathbb{R}^n$

$$\widehat{x} = x, \ \widehat{p} = -i\hbar\partial/\partial x_j.$$

What should then the operator $a_{rs}(\widehat{x}, \widehat{p})$ associated to the monomial $a_{rs}(x, p) = x^r p^s$ be? Born and Jordan's answer was

$$a_{rs}(\widehat{x}, \widehat{p}) = \frac{1}{s+1} \sum_{k=0}^{s} \widehat{p}^{s-k} \widehat{x}^r \widehat{p}^k. \tag{2.8}$$

They subsequently addressed the multi-dimensional case in a joint work [2] with Heisenberg himself. In [8] we have analyzed in detail Born and Jordan's argument, and shown that their approach to Heisenberg's matrix mechanics becomes effective if and only if one uses the quantization rule (2.8) for monomials. Born and Jordan's derivation has actually been discussed by many authors (see for instance Fedak and Prentis [10], Castellani [3], Crehan [5]), but to the best of our knowledge none has taken up the logical need for the rule (2.8). Approximately at the same time Hermann Weyl had started to develop his ideas about how to quantize the observables of a physical system, and communicated them to Max Born and Pascual Jordan (see Scholz [17] for a historical account). His basic ideas of a group theoretical approach were published two years later [19, 20]. A very interesting novelty in Weyl's approach was that he proposed to associate to an observable of a physical system what we would call today a pseudo-differential operator in Weyl form. In fact, writing the observable as an inverse Fourier transform

$$a(x,p) = \int e^{i(ps+xt)} Fa(s,t) ds dt \tag{2.9}$$

he defined its operator analogue by the formal substitution $x \longrightarrow \widehat{x}$, $p \longrightarrow \widehat{p}$, which yields

$$a(\widehat{x},\widehat{p}) = \int e^{i(\widehat{p}s+\widehat{x}t)} Fa(s,t) ds dt; \tag{2.10}$$

this is essentially the modern definition of a pseudo-differential operator in terms of Heisenberg operators. Now, Weyl's theory immediately yields the symmetrized quantization rule

$$a(\widehat{x},\widehat{p}) = \frac{1}{2}(\widehat{x}\widehat{p} + \widehat{p}\widehat{x})$$

(as does Born Jordan's algebraic constructions) and one finds that more generally (McCoy [15], 1932)

$$a_{rs}(\widehat{x},\widehat{p}) = \frac{1}{2^s}\sum_{k=0}^{s}\binom{s}{k}\widehat{p}^{s-k}\widehat{x}^r\widehat{p}^k \tag{2.11}$$

for a monomial $a_{rs}(x,p) = x^r p^s$.

We now make an essential observation. It turns out that Weyl's quantization rule (2.11) for monomials is a particular case of the so-called "τ-ordering" introduced by Shubin [18]: for any real number τ one defines the operator

$$a_{rs}^{\tau}(\widehat{x},\widehat{p}) = \sum_{k=0}^{s}\binom{s}{k}(1-\tau)^k\tau^{s-k}\widehat{p}^{s-k}\widehat{x}^r\widehat{p}^k \tag{2.12}$$

which is identical to Weyl's prescription when one chooses $\tau = \frac{1}{2}$. When $\tau = 0$ one gets the "normal ordering" $\widehat{x}^r\widehat{p}^s$ familiar from the elementary theory of partial differential equations while $\tau = 1$ yields the "anti-normal ordering" $\widehat{p}^s\widehat{x}^r$. We now make the following fundamental observation: the Born–Jordan prescription (2.8) is obtained by averaging (2.12) on the interval [0, 1]. In fact, noting that

$$\int_0^1 (1-\tau)^k\tau^{s-k} d\tau = \frac{k!(s-k)!}{(s+1)!}$$

we get

$$\int_0^1 a_{rs}^{\tau}(\widehat{x}, \widehat{p})d\tau = \frac{1}{s+1}\sum_{k=0}^{s} \widehat{p}^{s-k}\widehat{x}^r \widehat{p}^k \tag{2.13}$$

which is precisely (2.8).

In physics as well as in mathematics, the question of a "good" choice of quantization is more than just academic. For instance, different choices may lead to different spectral properties. The following example is due to Crehan [5]. Consider the Hamiltonian function

$$H(z) = \tfrac{1}{2}(p^2 + x^2) + \lambda(p^2 + x^2)^3.$$

The term that gives an ordering problem is evidently $(p^2 + x^2)^3$; Crehan then shows that the most general quantization invariant under the substitution $(x, p) \longmapsto (p, -x)$ is

$$\widehat{H} = \frac{1}{2}(\widehat{p}^2 + \widehat{x}^2) + \lambda(\widehat{p}^2 + \widehat{x}^2)^3 + \lambda(3\alpha\hbar^2 - 4)(\widehat{p}^2 + \widehat{x}^2).$$

It is easy to see that the eigenfunctions of $\widehat{H}$ are those of the harmonic oscillator $\widehat{H}_0 = \frac{1}{2}(\widehat{p}^2 + \widehat{x}^2)$ (they are thus the Hermite functions) and do not depend on the choices of the parameters λ and α. However the corresponding eigenvalues do: they are the numbers

$$E_N = (N + \tfrac{1}{2})\hbar + \lambda\hbar(2N + 1)^3 + \lambda\hbar(2N + 1)(3\alpha\hbar^2 - 4)$$

for $N = 0, 1, 2, ...$, which clearly shows the dependence of the spectrum on the parameters α and λ, and hence of the chosen quantization. This example clearly shows that the choice of a quantization is not just an academic problem, but has deep consequences when one looks for the correct spectra in physics. There are more subtle issues associated with the choice of quantization, and these will be discussed later on in this book.

We note that the ordering problem for monomials is still not closed, as witnessed by recent research (see for instance Domingo and Galapon [9]).

2.3 What Is Quantization?

In physics "quantization" refers to a mathematical procedure designed to describe a quantum system using its formulation as a classical system. We have been loosely talking about "quantization" as a process which allows one to associate an operator acting on some function space to a function; the latter is supposed to represent a dynamical variable, for instance energy, or position, or momentum; for a detailed and interesting discussion of the historical development of quantization, see Mehra and Rechenberg's treatise [16]. In the case of monomials the approach seems to be more abstract, because we associate to expressions like $x^r p^s$ a formal product of operators $\widehat{x}$ and $\widehat{p}$. It would therefore be useful to have a solid working mathematical definition of the notion of quantization. Let us immediately note that there is no consensus in the literature about what a "good" definition should be. We are going to give below a definition of quantization which is rather minimalistic, but sufficient for our purposes (and probably also the most reasonable from a physical point of view). But let us first explain what properties a quantization cannot satisfy; this will give us the opportunity of debunking what we called "urban legends" in [8]. The first of these properties is the so-called Dirac rule: any quantization $a \leftrightarrow \mathrm{Op}(a)$ should satisfy the relation

$$[\mathrm{Op}(a), \mathrm{Op}(b)] = i\hbar \mathrm{Op}(\{a, b\}) \tag{2.14}$$

where $\{a, b\}$ is the Poisson bracket of the two observables a, b. It is however well-known (the Groenewold–van Hove theorem, see [11, 12]) that (2.14) cannot hold for polynomials with degree > 2. Kauffmann gives in [14] an excellent analysis of Dirac's correspondence, and in [3] Castellani analyzes the (non-)existence of quantization rules satisfying (2.14). The second quantization rule that cannot be satisfied is von Neumann's condition

$$\mathrm{Op}(a^N) = (\mathrm{Op}(a))^N. \tag{2.15}$$

In fact, Cohen [4] has proven that this condition would prohibit the existence of a quasi-probability distribution $\rho(x, p)$ satisfying the marginal conditions

$$\int \rho(x, p) d^n p = |\psi(x)|^2, \quad \int \rho(x, p) d^n x = |F\psi(p)|^2 \tag{2.16}$$

and the average value formula

$$\langle g(\mathrm{Op}(a))\psi|\psi\rangle = \int g(a)(x, p)\rho(x, p) dp dx. \tag{2.17}$$

This would, among other unwanted consequences, prohibit the existence of the Wigner distribution and of a Weyl type phase space quantum mechanics.

So, now that we know what a quantization cannot be, let us list a few properties we would like a quantization to have.

Let us denote by Class(n) the vector space of all (real or complex) functions defined on phase space $\mathbb{R}^{2n}$; we do not assume any particular smoothness property for the elements of Class(n) (which we call "observables", or "symbols"). We will denote by Quant(n) the complex vector space of all continuous linear operators $\widehat{A} : \mathcal{S}(\mathbb{R}^n) \longrightarrow \mathcal{S}'(\mathbb{R}^n)$. We call *quantization* any linear mapping

$$\mathrm{Op} : \mathrm{Class}(n) \longrightarrow \mathrm{Quant}(n)$$

having the following properties:

- **Triviality axiom**:

$$\mathrm{Op}(1) = I_{\mathrm{d}}, \quad \mathrm{Op}(x_j) = \widehat{x}_j, \quad \mathrm{Op}(p_j) = \widehat{p}_j$$

(I_{d} the identity operator);
- **Self-adjointness**: if $a = a(x, p)$ is real, then $\mathrm{Op}(a)$ is self-adjoint; more generally:

$$\mathrm{Op}(a^*) = \mathrm{Op}(a)^\dagger$$

where a^* is the complex conjugate of a.

These two first properties are well-known, and very "reasonable"; the third axiom seems a little bit artificial, but helps maintain a relatively small class of possible quantizations:

- **Reduced Dirac correspondence**:

$$[\widehat{x}_j, \mathrm{Op}(a)] = i\hbar\mathrm{Op}(\{x_j, a\}),$$
$$[\widehat{p}_j, \mathrm{Op}(a)] = i\hbar\mathrm{Op}(\{p_j, a\})$$

for every $a \in \mathrm{Class}(n)$ and $j = 1, \ldots, n$.

It turns out that, at least as far as monomials or polynomials are concerned, the property above allows one to give very explicit expressions for $\mathrm{Op}(a)$; in particular one can prove the existence of a function f such that $f(0)$ and

$$\mathrm{Op}(x^r p^s) = \sum_{j=0}^{\min(r,s)} f^{(j)}(0)\binom{s}{j}\binom{r}{j} j!\hbar^j \widehat{p}^{s-j}\widehat{x}^{r-j} \tag{2.18}$$

(see Domingo and Galapon [9]). This property makes it easy to connect quantization—in the general case—with the theory of the Cohen classes, which plays an essential role in phase space quantum mechanics (and in its cousin, time-frequency analysis). We will come back to this property in Chap. 3.

A quantization scheme satisfying these three properties is called by some authors a "generalized Weyl correspondence"; we will not use this terminology because it gives the impression that the Weyl correspondence plays a privileged and central role in quantization. While it is true that the Weyl correspondence is in a sense the simplest quantization scheme, and that other quantization schemes can be studied in terms of it, it is not necessarily the best one in physics, as our discussion below will show.

2.4 Motivation for Born–Jordan Quantization

As shortly argued above there are many reasons to believe that the Born–Jordan ordering, which leads to the Born–Jordan pseudo-differential calculus is the correct physical quantization scheme. We have shown in [7, 8] that the equivalence of the Schrödinger and Heisenberg pictures of quantum mechanics (which is taken for granted in quantum physics) requires that the " observables" be quantized using the Born–Jordan rule. In fact, close scrutiny of Born and Jordan's argument shows that their quantization rule (2.13) is not only sufficient, but also *necessary* for their definitions to be mathematically consistent.

In the Schrödinger picture of quantum mechanics (wave mechanics), the operators are constant (unless they are explicitly time-dependent), and the states evolve in time: $\psi(t) = U(t, t_0)\psi(t_0)$ where

$$U(t, t_0) = e^{iH_S(t-t_0)/\hbar} \tag{2.19}$$

is a family of unitary operators (the propagator); the time evolution of ψ is thus governed by Schrödinger's equation

$$i\hbar\frac{\partial\psi}{\partial t} = H_S\psi; \tag{2.20}$$

H_S is an operator associated with the classical Hamiltonian function H by some "quantization rule". In the Heisenberg picture (matrix mechanics), the state vectors are time-independent operators that incorporate a dependency on time, while an observable A_S in the Schrödinger picture becomes a time-dependent operator $A_{\mathcal{H}}(t)$ in the Heisenberg picture; this time dependence satisfies the Heisenberg equation

$$i\hbar\frac{dA_{\mathcal{H}}}{dt} = i\hbar\frac{\partial A_{\mathcal{H}}}{\partial t} + [A_{\mathcal{H}}, H_{\mathcal{H}}]. \tag{2.21}$$

Schrödinger [6] (and, independently, Eckart [5]) attempted to prove shortly after the publication of Heisenberg's result that wave mechanics and matrix mechanics were mathematically equivalent. Both proofs contained flaws, and one had to wait until von Neumann's [7] seminal work for a rigorous proof of the equivalence of both theories. We will not bother with the technical shortcomings of Schrödinger's and Eckart's approaches here, but rather focus on one, perhaps more fundamental, aspect which seems to have been overlooked in the literature. We observe that it is possible to go from the Heisenberg picture to the Schrödinger picture (and back) using the following simple argument: a ket

$$|\psi_S(t)\rangle = U(t, t_0)|\psi_S(t_0)\rangle \tag{2.22}$$

in the Schrödinger picture becomes, in the Heisenberg picture, the constant ket

$$|\psi_{\mathcal{H}}\rangle = U(t, t_0)^{\dagger}|\psi_{\mathcal{S}}(t)\rangle = |\psi_{\mathcal{S}}(t_0)\rangle \tag{2.23}$$

whereas an observable $A_{\mathcal{S}}$ becomes

$$A_{\mathcal{H}}(t) = U(t, t_0)^{\dagger} A_{\mathcal{S}} U(t, t_0); \tag{2.24}$$

in particular the Hamiltonian is

$$H_{\mathcal{H}}(t) = U(t, t_0)^{\dagger} H_{\mathcal{S}} U(t, t_0). \tag{2.25}$$

Taking $t = t_0$ this relation implies that $H_{\mathcal{H}}(t_0) = H_{\mathcal{S}}$; now in the Heisenberg picture energy is constant, so the Hamiltonian operator $H_{\mathcal{H}}(t)$ must be a constant of the motion. It follows that $H_{\mathcal{H}}(t) = H_{\mathcal{S}}$ for all times t and hence both operators $H_{\mathcal{H}}$ and $H_{\mathcal{S}}$ must be quantized using the *same* rules. A consequence of this property is that if we believe that Heisenberg's "matrix mechanics" is correct and is equivalent to Schrödinger's theory, then the Hamiltonian operator appearing in the Schrödinger equation (2.20) *must* be quantized using the Born–Jordan rule, and not, as is usual in quantum mechanics, the Weyl quantization rule.

Now, why should we then choose the Born–Jordan quantization scheme, and not, for instance, the Weyl correspondence? It turns out that Born and Jordan's argument only works if one uses the quantization scheme that they proposed. We have explained this in detail in [8]; for completeness we reproduce here the argument (with some simplifications). A close scrutiny of the arguments in Born and Jordan [1] and its follow-up [2] by Born et al. shows that the key to their approach lies in the differentiation rule for products of non-commuting variables. They actually give two definitions, and prove thereafter that both coincide if and only if one makes an essential assumption on the ordering of the quantization of monomials. The first definition is algebraic: if

$$y = \prod_{m=1}^{s} y_{\ell_m} = y_{\ell_1} y_{\ell_2} \cdots y_{\ell_s} \tag{2.26}$$

is a product of non-commuting variables y_ℓ then, if $k \in \{\ell_1, \ell_2, ..., \ell_s\}$, the derivative of y with respect to y_k is given by what they call a "differential quotient of first type":

$$\left(\frac{\partial y}{\partial y_k}\right)_1 = \sum_{r=1}^{s} \delta_{\ell_r k} \prod_{m=r+1}^{s} y_{\ell_m} \prod_{m=1}^{r-1} y_{\ell_m} \tag{2.27}$$

($\delta_{\ell_r k}$ the Kronecker delta). In words: pick a factor x_k in (2.26) and form the product of all the following factors, and thereafter the product of the preceding factors (in that order). When y is a monomial $\widehat{p}^s \widehat{x}^r$ this rule yields

$$\left(\frac{\partial}{\partial \widehat{p}}(\widehat{p}^s\widehat{x}^r)\right)_1 = \sum_{\ell=0}^{s-1} \widehat{p}^{s-1-\ell}\widehat{x}^r\widehat{p}^\ell \tag{2.28}$$

$$\left(\frac{\partial}{\partial \widehat{x}}(\widehat{p}^s\widehat{x}^r)\right)_1 = \sum_{j=0}^{r-1} \widehat{x}^{r-1-j}\widehat{p}^s\widehat{x}^j. \tag{2.29}$$

The second definition (explicitly given in formula (3) of [2]) is similar to that of an ordinary partial derivative:

$$\left(\frac{\partial y}{\partial y_k}\right)_2 = \lim_{\alpha\to 0}\frac{f(\cdots, y_k+\alpha, \cdots)}{\alpha}. \tag{2.30}$$

With this definition formulas (2.28) and (2.29) become

$$\left(\frac{\partial}{\partial \widehat{p}}(\widehat{p}^s\widehat{x}^r)\right)_2 = s\widehat{p}^{s-1}\widehat{x}^r$$
$$\left(\frac{\partial}{\partial \widehat{x}}(\widehat{p}^s\widehat{x}^r)\right)_2 = r\widehat{p}^s\widehat{x}^{r-1}.$$

Their next step consists in identifying both notions of partial derivative; more specifically they want that the quantization $\widehat{H}$ (still to be defined) of a Hamiltonian function satisfies the equalities

$$\left(\frac{\partial \widehat{H}}{\partial \widehat{p}}\right)_1 = \left(\frac{\partial \widehat{H}}{\partial \widehat{p}}\right)_2, \left(\frac{\partial \widehat{H}}{\partial \widehat{x}}\right)_1 = \left(\frac{\partial \widehat{H}}{\partial \widehat{x}}\right)_2. \tag{2.31}$$

They thereafter show quite explicitly (in the footnote (1) of [2]) that these equations hold if the quantization $\widehat{H}$ of $H = p^s x^r$ is the self-adjoint operator given by

$$\widehat{H} = \frac{1}{r+1}\sum_{j=0}^{r} \widehat{x}^{r-j}\widehat{p}^s\widehat{x}^j = \frac{1}{s+1}\sum_{\ell=0}^{s} \widehat{p}^{s-\ell}\widehat{x}^r\widehat{p}^\ell. \tag{2.32}$$

They do not, however, show that it is the only possibility leading to a self-adjoint operator $\widehat{H}$. This is however the case, as we have shown in [8].

To be complete, let us explain why Born and Jordan needed these constructions. They assumed that the equations of motion for $\widehat{p}$ and $\widehat{x}$ are formally the same as in Hamiltonian mechanics, namely

$$\frac{d\widehat{x}}{dt} = \frac{\partial \widehat{H}}{\partial \widehat{p}}, \quad \frac{d\widehat{p}}{dt} = -\frac{\partial \widehat{H}}{\partial \widehat{x}}. \tag{2.33}$$

Pursuing this classical analogy, they require in addition that the Hamilton equations, written in terms of Poisson brackets

$$\frac{dx}{dt} = \{x, H\}, \quad \frac{dp}{dt} = \{p, H\}$$

should be replaced with the operator relations

$$\frac{d\widehat{x}}{dt} = [\widehat{x}, \widehat{H}], \quad \frac{d\widehat{p}}{dt} = [\widehat{p}, \widehat{H}];$$

to be consistent with the Hamilton equations (2.33) one must thus have

$$[\widehat{x}, \widehat{H}] = i\hbar\frac{\partial\widehat{H}}{\partial\widehat{p}}, \quad [\widehat{p}, \widehat{H}] = i\hbar\frac{\partial\widehat{H}}{\partial\widehat{x}}. \tag{2.34}$$

This last step in Born and Jordan's construction also requires that the operator $\widehat{H}$ must be given by the rule (2.32) above.

References

1. M. Born, P. Jordan, Zur Quantenmechanik. Zeits. Physik **34**, 858–888 (1925)
2. M. Born, W. Heisenberg, P. Jordan, Zur Quantenmechanik II. Z. Physik **35**, 557–615 (1925); English translation in: M. Jammer, *The Conceptual Development of Quantum Mechanics* (McGraw-Hill, New York, 1966); 2nd edn. (American Institute of Physics, New York, 1989)
3. L. Castellani, Quantization Rules and Dirac's Correspondence. Il Nuovo Cimento **48A**(3), 359–368 (1978)
4. L. Cohen, Generalized phase-space distribution functions. J. Math. Phys. **7**, 781–786 (1966)
5. P. Crehan, The parametrisation of quantisation rules equivalent to operator orderings, and the effect of different rules on the physical spectrum. J. Phys. A: Math. Gen. 811–822 (1989)
6. M. de Gosson, Symplectic covariance properties for Shubin and Born-Jordan pseudo-differential operators. Trans. Amer. Math. Soc. **365**(6), 3287–3307 (2013)
7. M. de Gosson, Born-Jordan quantization and the equivalence of the Schrödinger and Heisenberg pictures. Found. Phys. **44**(10), 1096–1106 (2014)
8. M. de Gosson, From Weyl to Born–Jordan quantization: the Schrödinger representation revisited. Phys. Rep. [in print] 2015
9. H.B. Domingo, E.A. Galapon, Generalized Weyl transform for operator ordering: polynomial functions in phase space. J. Math. Phys. **56**, 022104 (2015)
10. W.A. Fedak, J.J. Prentis, The 1925 Born and Jordan paper "On quantum mechanics". Am. J. Phys. **77**(2), 128–139 (2009)
11. M.J. Gotay, On the Groenewold-Van Hove problem for R^{2n}. J. Math. Phys. **40**(4), 2107–2116 (1999)
12. M.J. Gotay, H.B. Grundling, G.M. Tuynman, Obstruction results in quantization theory. J. Nonlinear Sci. **6**, 469–498 (1996)
13. W. Heisenberg, Über quantentheoretische Umdeutung kinematischer und mechanischer Beziehungen. Z. Physik **33**, 879–893 (1925)
14. S.K. Kauffmann, Unambiguous quantization from the maximum classical correspondence that is self-consistent: the slightly stronger canonical commutation rule dirac missed. Found. Phys. **41**(5), 805–819 (2011)

15. N.H. McCoy, On the function in quantum mechanics which corresponds to a given function in classical mechanics. Proc. Natl. Acad. Sci. U.S.A. **18**(11), 674–676 (1932)
16. J. Mehra, H. Rechenberg, *The Historical Development of Quantum Theory*, vol. 1. The Quantum Theory of Planck, Einstein, Bohr, and Sommerfeld: Its Foundation and the Rise of Its difficulties (Springer-Verlag, Berlin, 1980)
17. E. Scholz, *Weyl Entering the 'new' Quantum Mechanics Discourse*, ed. by C. Joas, C. Lehner, J. Renn. HQ-1: Conference on the History of Quantum Physics (Berlin, July 2–6, 2007). Preprint MPI History of Science, 350 vol. II (Berlin, 2007)
18. M.A. Shubin, *Pseudodifferential Operators and Spectral Theory*. Springer-Verlag, (1987) [original Russian edition in Nauka, Moskva (1978)]
19. H. Weyl, Quantenmechanik und Gruppentheorie. Zeitschrift für Physik, 46 (1927)
20. H. Weyl, *The Theory of Groups and Quantum Mechanics*, translated from the 2nd German edition by H.P. Robertson (Dutten, New York, 1931)

Chapter 3
Quantization of Monomials

In this chapter we begin by collecting some facts about the quantization of monomials and polynomials, with a particular emphasis on the Weyl and Born–Jordan schemes. We will also consider a non-standard rule, namely Shubin's τ-ordering, which is the key to the definition of Born–Jordan quantization for general observables. We thereafter propose a simple but efficient definition of quantization. That definition, due to Domingo and Galapon, is simple, because it relies on only three axioms (in addition to the requirement of linearity), and it is efficient because it is easy to generalize to the case of arbitrary observables, as we will see in the forthcoming chapters.

3.1 Polynomial Algebras

3.1.1 General Considerations and Notation

The first contribution to the topic of monomial quantization is Born and Jordan's seminal paper [2]; the list of papers that has followed this work is huge, and still growing (this book is not an exception!). Here is a short list of contributions, which is by no means exhaustive: Agarwal and Wolf [1], Crehan [3], Domingo and Galapon [5], Kerner and Sutcliffe [7], McCoy [8], Mehta [9], Misra and Shankara [10], Niederle and Tolar [11], Przanowski and Tosiek [12], Shewell [14].

In what follows x and p denote two indeterminates and $\mathbb{C}[x, p]$ the polynomial ring they generate: it consists of all finite formal sums $a = \sum_{r,s} \alpha_{rs} x^r p^s$ where the coefficients α_{rs} are complex numbers; it is assumed that $x^r p^s = p^s x^r$ hence $\mathbb{C}[x, p]$ is a commutative ring. We identify $\mathbb{C}[x, p]$ with the corresponding ring of polynomial functions. We will denote by $\mathbb{C}[\widehat{x}, \widehat{p}]$ the corresponding Weyl algebra: it is the

M.A. de Gosson, *Born–Jordan Quantization*, Fundamental Theories of Physics 182, DOI 10.1007/978-3-319-27902-2_3

complex (non-commutative) unital algebra[1] generated by $\widehat{x}$ and $\widehat{p}$, two indeterminates satisfying Born's canonical commutation relation (CCR), written formally as

$$\widehat{x}\widehat{p} - \widehat{p}\widehat{x} = i\hbar 1$$

where 1 is the unit of $\mathbb{C}[\widehat{x}, \widehat{p}]$; we will commonly abuse notation by writing $i\hbar 1 \equiv i\hbar$; see Gelfand and Fairlie [6] for a detailed study of the Weyl algebra $\mathbb{C}[\widehat{x}, \widehat{p}]$.

The following commutation formulas in $\mathbb{C}[\widehat{x}, \widehat{p}]$ are easily proven by induction on the integers r and s. Using these formulas it is easy to see that every $\widehat{A} \in \mathbb{C}[\widehat{x}, \widehat{p}]$ can be written in any of the two forms below:

$$\widehat{A} = \sum_{r+s\leq m} \alpha_{rs}\widehat{x}^r \widehat{p}^s = \sum_{r+s\leq m} \beta_{rs}\widehat{p}^s\widehat{x}^r.$$

In standard quantum mechanics it is usual to choose for $\widehat{x}$ and $\widehat{p}$ the operators defined by

$$\widehat{x}\psi = x\psi, \widehat{p} = -i\hbar\frac{\partial\psi}{\partial x}$$

where ψ is a differentiable function on the real line. This is not the only possible choice; for instance in the theory of the phase space Schrödinger equations (and in its variant, deformation quantization), one often uses instead the so-called "Bopp shifts" $\widetilde{x}$ and $\widetilde{p}$, which are defined by the relations

$$\widetilde{x}\psi = \left(x + \frac{1}{2}i\hbar\frac{\partial}{\partial p}\right)\psi, \widetilde{p}\psi = \left(p - \frac{1}{2}i\hbar\frac{\partial}{\partial x}\right)\psi$$

where ψ is this time a differentiable function on phase space $\mathbb{R}^2$. It is immediate to verify that $[\widetilde{x}, \widetilde{p}] = i\hbar 1$.

Let us begin by giving a very rough definition of quantization: a *quantization* of $\mathbb{C}[x, p]$ is a linear mapping

$$\mathrm{Op} : \mathbb{C}[x, p] \longrightarrow W[\widehat{x}, \widehat{p}]$$

associating to each polynomial $a(x, p) \in \mathbb{C}[x, p]$ with *real* coefficients a self-adjoint element $\widehat{A} = a(\widehat{x}, \widehat{p}) \in \mathbb{C}[\widehat{x}, \widehat{p}]$. Self-adjointness is taken here in the algebraic sense: if

$$\widehat{A} = \sum_{r+s\leq m} \alpha_{rs}\widehat{x}^r \widehat{p}^s \in \mathbb{C}[\widehat{x}, \widehat{p}]$$

[1] It can be viewed, if one wants, as the universal enveloping algebra of the Heisenberg Lie algebra.

then the adjoint $\widehat{A}^\dagger$ of $\widehat{A}$ is defined by

$$\widehat{A}^\dagger = \sum_{r+s\leq m} \alpha^*_{rs}\widehat{p}^s\widehat{x}^r \in \mathbb{C}[\widehat{x},\widehat{p}];$$

the fact that $\widehat{A}^\dagger \in \mathbb{C}[\widehat{x},\widehat{p}]$ follows either by repeated use of the CCR (see the commutation formula (3.3) below): we have

$$\widehat{A}^\dagger = \sum_{r',s'} \beta_{r's'}\widehat{x}^{r'}\widehat{p}^{s'}$$

where the new coefficients $\beta_{r's'}$ can be stepwise determined from the α_{rs}.

3.1.2 *Commutation Relations*

Using Born's canonical commutation relation

$$[\widehat{x},\widehat{p}] = \widehat{x}\widehat{p} - \widehat{p}\widehat{x} = i\hbar$$

it is easy to prove by induction on the integers r and s various commutation relations. Here are a few of them:

$$[\widehat{x}^r,\widehat{p}^s] = si\hbar\sum_{\ell=0}^{r-1}\widehat{x}^{r-1-\ell}\widehat{p}^{s-1}\widehat{x}^{\ell} \tag{3.1}$$

$$[\widehat{x}^r,\widehat{p}^s] = ri\hbar\sum_{j=0}^{s-1}\widehat{p}^{s-1-j}\widehat{x}^{r-1}\widehat{p}^{j} \tag{3.2}$$

$$[\widehat{x}^r,\widehat{p}^s] = \sum_{k=1}^{\min(r,s)}(i\hbar)^k k!\binom{r}{k}\binom{s}{k}\widehat{p}^{s-k}\widehat{x}^{r-k} \tag{3.3}$$

$$[\widehat{x}^r,\widehat{p}^s] = -\sum_{k=1}^{\min(r,s)}(-i\hbar)^k k!\binom{r}{k}\binom{s}{k}\widehat{x}^{r-k}\widehat{p}^{s-k}. \tag{3.4}$$

Notice that the relations (3.3) and (3.4) are obtained from each other by taking adjoints and observing that the commutator is antisymmetric: we have $[\widehat{x}^r,\widehat{p}^s]^\dagger = [\widehat{p}^s,\widehat{x}^r]$.

3.2 Some Common Orderings

We briefly discuss here the Weyl and Born–Jordan quantizations. The latter, given by

$$\mathrm{Op}_{\mathrm{BJ}}(p^s x^r) = \frac{1}{s+1}\sum_{\ell=0}^{s} \widehat{p}^{s-\ell}\widehat{x}^r \widehat{p}^\ell$$

is the *equally weighted average* of all the possible operator orderings, as opposed to the Weyl correspondence

$$\mathrm{Op}_{\mathrm{W}}(x^r p^s) = \frac{1}{2^s}\sum_{\ell=0}^{s} \binom{s}{\ell} \widehat{p}^{s-\ell}\widehat{x}^r \widehat{p}^\ell$$

or the symmetric ordering (Rivier [13])

$$\mathrm{Op}_{\mathrm{sym}}(x^r p^s) = \frac{1}{2}(\widehat{p}^s\widehat{x}^r + \widehat{x}^r\widehat{p}^s)$$

which is the most *symmetrical* rule. All three rules coincide when $s + r \leq 2$, but they are different as soon as $s \geq 2$ and $r \geq 2$.

3.2.1 Weyl Ordering

By definition the Weyl quantization (or correspondence) Op_{W} associates to a monomial $x^r p^s$ the expression (McCoy [8])

$$\mathrm{Op}_{\mathrm{W}}(x^r p^s) = \frac{1}{2^s}\sum_{\ell=0}^{s} \binom{s}{\ell} \widehat{p}^{s-\ell}\widehat{x}^r \widehat{p}^\ell; \tag{3.5}$$

equivalently, using the commutation relations (3.1)–(3.4),

$$\mathrm{Op}_{\mathrm{W}}(x^r p^s) = \frac{1}{2^r}\sum_{\ell=0}^{r} \binom{s}{\ell} \widehat{x}^{\ell}\widehat{p}^s \widehat{x}^{r-\ell}.$$

Example 1 We have $\mathrm{Op}_{\mathrm{W}}(xp) = \frac{1}{2}(\widehat{x}\widehat{p} + \widehat{p}\widehat{x})$ and

$$\mathrm{Op}_{\mathrm{W}}(x^2 p^2) = \tfrac{1}{4}(\widehat{x}^2\widehat{p}^2 + 2\widehat{x}\widehat{p}^2\widehat{x} + \widehat{p}^2\widehat{x}^2)$$

or, equivalently,

$$\mathrm{Op}_{\mathrm{W}}(x^2 p^2) = \tfrac{1}{4}(\widehat{p}^2\widehat{x}^2 + 2\widehat{p}\widehat{x}^2\widehat{p} + \widehat{x}^2\widehat{p}^2)$$

(the somewhat counterintuitive equality $\widehat{x}\widehat{p}^2\widehat{x} = \widehat{p}\widehat{x}^2\widehat{p}$ immediately follows from the commutation relation $[\widehat{x}, \widehat{p}] = i\hbar$).

The definition above immediately makes apparent the self-adjointness of the Weyl quantization of real polynomials: we have

$$\mathrm{Op_W}(x^r p^s)^\dagger = \frac{1}{2^s}\sum_{\ell=0}^{s}\binom{s}{\ell}\widehat{p}^\ell\widehat{x}^r\widehat{p}^{s-\ell} = \mathrm{Op_W}(x^r p^s),$$

and this formula immediately extends to polynomials by linearity.

Proposition 2 *Let $x^r p^s \in \mathbb{C}[x,p]$; we have the "normal" and "anti-normal" expansions*

$$\mathrm{Op_W}(x^r p^s) = \sum_{\ell=0}^{\min(r,s)}(-i\hbar)^\ell\binom{s}{\ell}\binom{r}{\ell}\frac{\ell!}{2^\ell}\widehat{x}^{r-\ell}\widehat{p}^{s-\ell}; \tag{3.6}$$

$$\mathrm{Op_W}(x^r p^s) = \sum_{\ell=0}^{\min(r,s)}(i\hbar)^\ell\binom{s}{\ell}\binom{r}{\ell}\frac{\ell!}{2^\ell}\widehat{p}^{s-\ell}\widehat{x}^{r-\ell}. \tag{3.7}$$

Proof Using the formulas (3.1)–(3.4) it is straightforward to show that $\mathrm{Op_W}(x^r p^s)$ can be written in any of the two forms above; notice that both formulas are easily seen to be equivalent if one takes into account the self-adjointness property of $\mathrm{Op_W}(x^r p^s)$ since the adjoint of $\widehat{x}^{r-\ell}\widehat{p}^{s-\ell}$ is precisely $\widehat{p}^{s-\ell}\widehat{x}^{r-\ell}$. ∎

Example 3 While by definition we have

$$\mathrm{Op_W}(x^2p^2) = \tfrac{1}{4}(\widehat{x}^2\widehat{p}^2 + 2\widehat{x}\widehat{p}^2\widehat{x} + \widehat{p}^2\widehat{x}^2)$$

using formula (3.6) with $r = s = 2$ this is the same thing as

$$\mathrm{Op_W}(x^2p^2) = \widehat{x}^2\widehat{p}^2 - 2i\hbar\widehat{x}\widehat{p} - \tfrac{1}{2}\hbar^2 \tag{3.8}$$

as can be directly checked using by repeated use of the commutation relation $[\widehat{x},\widehat{p}] = i\hbar$; similarly (3.7) yields

$$\mathrm{Op_W}(x^2p^2) = \widehat{p}^2\widehat{x}^2 + 2i\hbar\widehat{p}\widehat{x} - \tfrac{1}{2}\hbar^2. \tag{3.9}$$

The assumption of linearity allows us to construct $\mathrm{Op_W}(a)$ for arbitrary polynomials $a \in \mathbb{C}[x,p]$: if

$$a(x,p) = \sum_{r+s\leq m}\alpha_{rs}x^r p^s$$

then we have

$$\mathrm{Op_W}(a) = \sum_{r+s\leq m}\alpha_{rs}\mathrm{Op_W}(x^r p^s).$$

3.2.2 Born–Jordan Ordering

By definition, the Born–Jordan ordering of monomials is given by

$$\mathrm{Op_{BJ}}(x^r p^s) = \frac{1}{s+1}\sum_{\ell=0}^{s} \widehat{p}^{s-\ell}\widehat{x}^r\widehat{p}^{\ell} \tag{3.10}$$

or, equivalently,

$$\mathrm{Op_{BJ}}(x^r p^s) = \frac{1}{r+1}\sum_{j=0}^{r} \widehat{x}^{r-j}\widehat{p}^s\widehat{x}^{j}. \tag{3.11}$$

The equivalence of both definitions follows from the commutation relations (3.1)–(3.4).

Since the adjoint of $\widehat{p}^{s-\ell}\widehat{x}^r\widehat{p}^{\ell}$ is $\widehat{p}^{\ell}\widehat{x}^r\widehat{p}^{s-\ell}$ we have $\mathrm{Op_{BJ}}(p^s x^r)^\dagger = \mathrm{Op_{BJ}}(p^s x^r)$, hence the Born–Jordan rule is, as is the Weyl rule, a physically acceptable quantization scheme for monomials. It extends to polynomials by linearity: if

$$a(x,p) = \sum_{r+s\leq m} \alpha_{rs} x^r p^s$$

is an arbitrary element of $\mathbb{C}[x,p]$ then we define $\mathrm{Op_{BJ}}(a) \in \mathbb{C}[\widehat{x},\widehat{p}]$ by

$$\mathrm{Op_{BJ}}(a) = \sum_{r+s\leq m} \alpha_{rs}\mathrm{Op_{BJ}}(x^r p^s).$$

The following result gives the normal and anti-normal expansions in the Born–Jordan scheme:

Proposition 4 *Let $x^r p^s \in \mathbb{C}[x,p]$; we have*

$$\mathrm{Op_{BJ}}(x^r p^s) = \sum_{\ell=0}^{\min(r,s)} (-i\hbar)^{\ell}\binom{s}{\ell}\binom{r}{\ell}\frac{\ell!}{\ell+1}\widehat{x}^{r-\ell}\widehat{p}^{s-\ell} \tag{3.12}$$

$$\mathrm{Op_{BJ}}(x^r p^s) = \sum_{\ell=0}^{\min(r,s)} (i\hbar)^{\ell}\binom{s}{\ell}\binom{r}{\ell}\frac{\ell!}{\ell+1}\widehat{p}^{s-\ell}\widehat{x}^{r-\ell}. \tag{3.13}$$

Proof Similar to that of Proposition 2 in the Weyl case; also see [7]; observe that one toggles between both expressions by taking the adjoints. ■

Example 5 Setting $r = s = 2$ in (3.12) we get the normally ordered operator

$$\mathrm{Op_{BJ}}(x^2 p^2) = \widehat{x}^2\widehat{p}^2 - 2i\hbar\widehat{x}\widehat{p} - \tfrac{2}{3}\hbar^2; \tag{3.14}$$

it is different from the expression (3.8) giving the normally ordered Weyl quantization of $x^2 p^2$.

Observe that both formulas (3.6) and (3.12) are particular cases of the formula

$$\mathrm{Op}_{(\alpha_\ell)}(x^r p^s) = \sum_{\ell=0}^{\min(r,s)} (-i\hbar)^\ell \binom{s}{\ell}\binom{r}{\ell}\frac{\ell!}{\alpha_\ell}\widehat{x}^{r-\ell}\widehat{p}^{s-\ell} \tag{3.15}$$

where (α_ℓ) is a sequence of real numbers. For instance, the choice $\alpha_\ell = \ell+1$ leads to the Born–Jordan rule, and the choice $\alpha_\ell = 2^\ell$ leads to the Weyl rule. These formulas show explicitly that the Weyl and Born–Jordan rules differ as soon as $r \geq 2$ and $s \geq 2$ (cf. Dewey [4]). We will give a more explicit formula below.

3.2.3 *The Relation Between* $\mathrm{Op_W}(x^r p^s)$ *and* $\mathrm{Op_{BJ}}(x^r p^s)$

We shortly address here the following question: "can we (easily) find the Born–Jordan quantization of $x^r p^s$ knowing its Weyl quantization, and *vice versa*?". While nothing is really very simple when it comes to such combinatorial matters, the two following formulas have been proven by Domingo and Galapon [5]:

$$\mathrm{Op_{BJ}}(x^r p^s) = \sum_{j=0}^{\frac{1}{2}\min(r,s)} \frac{(i\hbar/2)^{2j}}{(2j+1)!}\frac{s!}{(s-2j)!}\frac{r!}{(r-2j)!}\mathrm{Op_W}(x^{r-2j}p^{s-2j})$$

and, conversely,

$$\mathrm{Op_W}(x^r p^s) = \sum_{j=0}^{\min(r,s)} \frac{(i\hbar)^j B_j(\frac{1}{2})}{j!}\frac{s!}{(s-j)!}\frac{r!}{(r-j)!}\mathrm{Op_{BJ}}(x^{r-2j}p^{s-2j});$$

in the second formula $B_j(\frac{1}{2})$ is the Bernoulli polynomial $B_j(x)$ evaluated at $x = \frac{1}{2}$; recall that $B_j(x)$ is defined by

$$\frac{te^{xt}}{e^t - 1} = \sum_{j=0}^{\infty} B_j(x)\frac{t^j}{j!}$$

or equivalently

$$B_j(x) = \sum_{k=0}^{j} \binom{j}{k} B_{j-k}x^k$$

where the B_{j-k} are the Bernoulli numbers.

Example 6 Choose $r = s = 2$. We have

$$\mathrm{Op_W}(x^2p^2) = \widehat{x}^2\widehat{p}^2 - 2i\hbar\widehat{x}\widehat{p} - \tfrac{1}{2}\hbar^2$$

and

$$\mathrm{Op_{BJ}}(x^rp^s) = \widehat{x}^2\widehat{p}^2 - 2i\hbar\widehat{x}\widehat{p} - \tfrac{2}{3}\hbar^2.$$

This example clearly shows that Born–Jordan and Weyl quantizations are different: subtracting the second equality from the first we obtain

$$\mathrm{Op_W}(x^2p^2) - \mathrm{Op_{BJ}}(x^rp^s) = -\tfrac{1}{6}\hbar^2.$$

3.2.4 Shubin's τ-Ordering

It turns out that the Weyl ordering is a particular case of what we call Shubin's τ-ordering, defined by

$$\mathrm{Op}_\tau(x^rp^s) = \sum_{\ell=0}^{s}\binom{s}{\ell}(1-\tau)^\ell\tau^{s-\ell}\widehat{p}^{s-\ell}\widehat{x}^r\widehat{p}^\ell \tag{3.16}$$

(τ is an arbitrary real number); equivalently

$$\mathrm{Op}_\tau(x^rp^s) = \sum_{\ell=0}^{r}\binom{r}{\ell}(1-\tau)^\ell\tau^{s-\ell}\widehat{x}^\ell\widehat{p}^s\widehat{x}^{r-\ell}. \tag{3.17}$$

Clearly, the choice $\tau = \frac{1}{2}$ immediately yields Weyl's rule; the choices $\tau = 0$ and $\tau = 1$ lead to the normal and antinormal rules

$$\mathrm{Op_N}(x^rp^s) = \widehat{x}^r\widehat{p}^s, \mathrm{Op_{AN}}(x^rp^s) = \widehat{p}^s\widehat{x}^r,$$

respectively. What is less obvious—and very interesting, indeed—is that if we integrate the right-hand side of (3.16) for τ going from 0 to 1, then we recover the Born–Jordan rule

$$\mathrm{Op_{BJ}}(x^rp^s) = \frac{1}{s+1}\sum_{\ell=0}^{s}\widehat{x}^\ell\widehat{p}^s\widehat{x}^{r-\ell};$$

this easily follows from the property

$$\int_0^1 \tau^{s-\ell}(1-\tau)^\ell d\tau = \frac{(s-\ell)!\ell!}{(s+1)!} \tag{3.18}$$

familiar from the theory of the beta function. This essential remark will allow us to define Born–Jordan quantization for arbitrary observables, by extending the formula

$$\mathrm{Op_{BJ}}(x^r p^s) = \int_0^1 \mathrm{Op}_\tau(x^r p^s) d\tau \tag{3.19}$$

to arbitrary Hamiltonian functions.

It should be remarked that the τ-rule is itself unphysical for $\tau \neq \frac{1}{2}$ because it does not associate to the real observable $p^s x^r$ a self-adjoint operator; in fact

$$\mathrm{Op}_\tau(x^r p^s)^\dagger = \mathrm{Op}_{1-\tau}(x^r p^s)$$

as immediately follows from (3.16) or (3.17). The Born–Jordan rule is, in contrast, physical since

$$\mathrm{Op_{BJ}}(x^r p^s)^\dagger = \mathrm{Op_{BJ}}(x^r p^s)$$

(this can be seen directly from its definition, or using the formula above).

3.3 General Quantization Axioms for Monomials

Quantization rules are a flourishing market; the monomial case is not an exception. Very recently Domingo and Galapon [5] have proposed a very simple set of rules and shown that it allows one to produce a general formula for the quantization of arbitrary polynomials in terms of a certain function χ of the variables x, p. It turns out the theory of Domingo and Galapon, in addition to being "minimalistic", is the gate to the quantization of arbitrary observables, as we will see in the forthcoming chapters. We follow here almost *verbatim* their presentation.

3.3.1 The Domingo–Galapon Formula

We begin by recalling the notion of quantization for monomials following [5]:

Definition 7 A quantization of $\mathbb{C}[x, p]$ is a linear mapping $\mathrm{Op} : \mathbb{C}[x, p] \longrightarrow \mathbb{C}[\widehat{x}, \widehat{p}]$ having the following properties:

(A1) $\mathrm{Op}(1) = 1$ (the identity), $\mathrm{Op}(x) = \widehat{x}$ and $\mathrm{Op}(p) = \widehat{p}$;
(A2) $\mathrm{Op}(a)$ is self-adjoint if $a \in \mathbb{C}[x, p]$ is a real polynomial;
(A3) The restricted Dirac correspondence

$$[\widehat{x}, \mathrm{Op}(a)] = i\hbar \mathrm{Op}(\{x, a\}) \tag{3.20}$$
$$[\widehat{p}, \mathrm{Op}(a)] = i\hbar \mathrm{Op}(\{p, a\}) \tag{3.21}$$

holds for every $a \in \mathbb{C}[x, p]$.

In the formulas above the curly brackets $\{\cdot, \cdot\}$ denote the Poisson bracket, familiar from classical mechanics:

$$\{a, b\} = \frac{\partial a}{\partial x}\frac{\partial b}{\partial p} - \frac{\partial b}{\partial x}\frac{\partial a}{\partial p};$$

hence the rules (3.20) and (3.21) mean that

$$\widehat{x}\mathrm{Op}(a) - \mathrm{Op}(a)\widehat{x} = i\hbar\mathrm{Op}\left(\frac{\partial a}{\partial p}\right)$$

and

$$\widehat{p}\mathrm{Op}(a) - \mathrm{Op}(a)\widehat{p} = i\hbar\mathrm{Op}\left(-\frac{\partial a}{\partial x}\right).$$

Domingo and Galapon [5] call such a quantization of $\mathbb{C}[x, p]$ a "generalized Weyl transform". The interest of their definition comes from the following result, which characterizes all quantizations of $\mathbb{C}[x, p]$ satisfying the axioms (A1)–(A2)–(A3):

Proposition 8 *Let* $\mathrm{Op} : \mathbb{C}[x, p] \longrightarrow \mathbb{C}[\widehat{x}, \widehat{p}]$ *be a quantization in the sense of Domingo and Galapon. Then there exists a real function* $\chi \in C^\infty(\mathbb{R}^2)$ *with* $\chi(0) = 1$ *such that*

$$\mathrm{Op}(x^r p^s) = \sum_{\ell=0}^{\min(r,s)} \hbar^\ell \ell! f^{(\ell)}(0)\binom{s}{\ell}\binom{r}{\ell}\widehat{p}^{s-\ell}\widehat{x}^{r-\ell} \tag{3.22}$$

where $f^{(\ell)}(0)$ *is the* ℓ*th derivative of the function* $f(x) = e^{ix/2}\chi(x)$ *at* $x = 0$.

Proof We are just going to indicate the main lines of the proof, and refer for details to [5]. One begins by writing $\widehat{A}_{sr} = \mathrm{Op}(x^s p^r)$ in polynomial form

$$\widehat{A}_{sr} = \sum_{\ell=0}^{\min(r,s)} g_{s,r,\ell}\hbar^\ell \widehat{p}^{s-\ell}\widehat{x}^{r-\ell}. \tag{3.23}$$

The adjoint of this operator is

$$\widehat{A}_{sr}^\dagger = \sum_{\ell=0}^{\min(r,s)} \sum_{k=0}^{\ell} g^*_{s,r,k} b^*_{s-k,r-k,\ell-k}\hbar^\ell \widehat{p}^{s-\ell}\widehat{x}^{r-\ell}$$

the coefficients $b_{s,r,\ell}$ being given by

$$b_{s,r,\ell} = \frac{(-1)^\ell s!r!/\ell!}{(s-\ell)!(r-\ell)!} \text{ for } \ell \leq \min(r, s)$$

and $b_{s,r,\ell} = 0$ for $\ell > \min(r, s)$. Since $x^s p^r$ is real we have $\widehat{A}_{sr}^\dagger = \widehat{A}_{sr}$ and this condition implies that

$$g_{s,r,\ell} = \sum_{j=0}^{\ell} g_{s,r,j}^* b_{s-j,r-j,\ell-j}^*. \tag{3.24}$$

Applying axiom (A3) in the definition above yields the equalities

$$[\widehat{x}, \widehat{A}_{sr}] = i\hbar s \widehat{A}_{s-1,r}, \; [\widehat{p}, \widehat{A}_{sr}] = -i\hbar r \widehat{A}_{s,r-1};$$

writing

$$\widehat{A}_{s-1,r} = \sum_{\ell=0}^{\min(s-1,r)} g_{s-1,r,\ell} \hbar^\ell \widehat{p}^{s-1-\ell} \widehat{x}^{r-\ell}$$

$$\widehat{A}_{s,r-1} = \sum_{\ell=0}^{\min(s,r-1)} g_{s,r-1,\ell} \hbar^\ell \widehat{p}^{s-\ell} \widehat{x}^{r-1-\ell}$$

this leads to the conditions

$$g_{s-1,r,\ell} = \frac{s-\ell}{s} g_{s,r,\ell}, \; g_{s,r-1,\ell} = \frac{r-\ell}{r} g_{s,r,\ell}.$$

Combining these two expressions we get

$$g_{s,r,\ell} = \binom{s}{\ell}\binom{r}{\ell} g_{\ell,\ell,\ell} \tag{3.25}$$

and substituting this expression in (3.24) gives the recurrence relation

$$\frac{g_{\ell,\ell,\ell}}{\ell!\ell!} = \sum_{k=0}^{\ell} \frac{g_{k,k,k}^*}{k!k!} \frac{i^{\ell-k}}{(\ell-k)!} \text{ for } \ell \leq \min(r, s).$$

Now comes the crucial step: define a function f by the series

$$f(x) = \sum_{\ell=0}^{\infty} \frac{g_{\ell,\ell,\ell}}{\ell!} \frac{x^\ell}{\ell!};$$

since $g_{\ell,\ell,\ell} = \ell! f^{(\ell)}(0)$ the recurrence formula above implies $e^{-ix/2} f(x) = e^{ix/2} f(x)^*$ hence the function $\chi(x) = e^{-ix/2} f(x)$ is real; moreover $\chi(0) = g_{0,0,0} = 1$. Formula (3.22) follows, inserting (3.25) in the expression (3.23) for $\widehat{A}_{sr}$. ■

Here are two examples.

Example 9 **Weyl ordering**. Assume first that the function χ is the identity: $\chi(x) = 1$ for all $x \in \mathbb{R}$. Then $f(x) = e^{ix/2}$ and $f^{(j)}(0) = (i/2)^j$. Formula (3.22) becomes in this case

$$\mathrm{Op}(x^r p^s) = \sum_{j=0}^{\min(r,s)} j! \binom{s}{j} \binom{r}{j} \frac{(i\hbar)^j}{2^j} \widehat{p}^{s-j} \widehat{x}^{r-j}$$

hence $\mathrm{Op}(x^r p^s) = \mathrm{Op_W}(x^r p^s)$ in view of formula (3.7).

Recall that the cardinal sine function sinc is defined by $\mathrm{sinc}(u) = (\sin u)/u$ for $u \neq 0$ and $\mathrm{sinc}(0) = 1$.

Example 10 **Born–Jordan ordering**. Assume now $\chi(x) = \mathrm{sinc}(x/2)$; then $f(x) = e^{ix/2}\,\mathrm{sinc}(x/2)$, and we have $f^{(\ell)}(0) = i^\ell/(\ell+1)$. Hence

$$\mathrm{Op}(x^r p^s) = \sum_{\ell=0}^{\min(r,s)} (i\hbar)^\ell \binom{s}{\ell} \binom{r}{\ell} \frac{\ell!}{\ell+1} \widehat{p}^{s-\ell} \widehat{x}^{r-\ell}$$

so that $\mathrm{Op}(x^r p^s) = \mathrm{Op_{BJ}}(x^r p^s)$ (formula (3.13).

The definition of quantization given above is very simple; in many texts one adds supplementary restrictive assumptions, see for instance Niederle and Tolar [11] or Przanowski and Tosiek [12]. We will not discuss such conditions here, because they are actually unnecessary complications of the theory.

We also note that it is often required in texts on quantization that one should in some way recover the classical observable by taking the "limit" $\hbar \to 0$. This is automatically satisfied if one uses the Domingo–Galapon formula (3.22) which implies that

$$\lim_{\hbar \to 0} \mathrm{Op}(x^r p^s) = \widehat{p}^s \widehat{x}^r$$

which can be written as

$$\lim_{\hbar \to 0} \mathrm{Op}(x^r p^s)\big|_{\widehat{x}=x,\, \widehat{p}=p} = x^r p^s; \tag{3.26}$$

but this notation is of course, mathematically speaking, rather formal, to say the least.

References

1. G.S. Agarwal, K.B. Wolf, Calculus for functions of noncommuting operators and general phase-space methods in quantum mechanics I. Mapping theorems and ordering of functions of noncommuting operators. Phys. Rev. D. **2**(10), 2161–2164 (1970)
2. M. Born, P. Jordan, Zur Quantenmechanik. Zeits. Physik **34**, 858–888 (1925)

3. P. Crehan, The parametrisation of quantisation rules equivalent to operator orderings, and the effect of different rules on the physical spectrum, J. Phys. A: Math. Gen. 811–822 (1989)
4. T.G. Dewey, Numerical mathematics of Feynman path integrals and the operator ordering problem. Phys. Rev. A **42**(1), 32–37 (1990)
5. H.B. Domingo, E.A. Galapon, Generalized Weyl transform for operator ordering: polynomial functions in phase space. J. Math. Phys. **56**, 022194 (2015)
6. I.M. Gelfand, D.B. Fairlie, The algebra of Weyl symmetrized polynomials and its quantum extension. Comm. Math. Phys. **136**(3), 487–499 (1991)
7. E.H. Kerner, W.G. Sutcliffe, Unique Hamiltonian operators via Feynman path integrals. J. Math. Phys. **11**(2), 391–393 (1970)
8. N.H. McCoy, On the function in quantum mechanics which corresponds to a given function in classical mechanics. Proc. Natl. Acad. Sci. U.S.A. **18**(11), 674–676 (1932)
9. C.L. Mehta, Phase-space formulation of the dynamics of canonical variables. J. Math. Phys. **5**(1), 677–686 (1963)
10. S.P. Misra, T.S. Shankara, Semiclassical and quantum description. J. Math. Phys. **9**(2), 299–304 (1968)
11. J. Niederle, J. Tolar, Quantization as mapping and as deformation. Czech. J. Phys. **B29**, 1358–1368 (1979)
12. M. Przanowski, J. Tosiek, Weyl–Wigner–Moyal formalism. I. Operator ordering. Acta. Phys. Pol. B **26**, 1703–1716 (1995)
13. D.C. Rivier, On a one-to-one correspondence between infinitesimal canonical transformations and infinitesimal unitary transformations. Phys. Rev. **83**(4), 862–863 (1951)
14. J.R. Shewell, On the formation of quantum-mechanical operators. Am. J. Phys. **27**, 16–21 (1959)

Chapter 4
Basic Hamiltonian Mechanics

In order to discuss wave mechanics and the Schrödinger equation in the next chapter we need some basic aspects of classical mechanics in its Hamiltonian formulation. For detailed accounts of Hamiltonian mechanics we refer to Abraham and Marsden [1], Arnol'd [2], Calkin [4], Goldstein [8], de Gosson [6, 7], Synge and Truesdell [12]. We begin by briefly discussing Hamilton's equations of motion, and we thereafter review the fundamental notions of generating function and Hamilton's two-point characteristic function, which are both related to *action*. In the last section we address the topic of short-time approximations to these functions; this topic is usually ignored in the literature, a notable exception being the paper [9] by Makri and Miller, whose results were independently later rediscovered in [7] using a different method. We thereafter extend these results to the case of arbitrary Hamiltonian functions.

4.1 Hamiltonian Dynamics

A matter of terminology and notation: we will comply with the traditional denomination "canonical transformations" for phase space diffeomorphisms which respect the symplectic structure determined by the matrix $J = \begin{pmatrix} 0_{n\times n} & I_{n\times n} \\ -I_{n\times n} & 0_{n\times n} \end{pmatrix}$. We will occasionally also use the synonym " symplectomorphism", which is more common in mathematical texts. Beware: in some physics texts a "canonical transformation" has the slightly more general meaning of a transformation which preserves the form of Hamilton's equations (see Arnol'd [2], Sect. 45, Footnote 76).

Let H be a real-valued function; we assume for convenience that $C^\infty(\mathbb{R}^{2n} \times \mathbb{R})$, but most of what follows remains true if we assume less stringent differentiability conditions (see Abraham and Marsden [1] for a discussion of sufficient smoothness requirements). The $2n$ ordinary differential equations

M.A. de Gosson, *Born–Jordan Quantization*, Fundamental Theories of Physics 182, DOI 10.1007/978-3-319-27902-2_4

$$\frac{dx_j}{dt} = \frac{\partial H}{\partial p_j}(x_1, \ldots, x_n, p_1, \ldots, p_n, t) \tag{4.1}$$

$$\frac{dp_j}{dt} = -\frac{\partial H}{\partial x_j}(x_1, \ldots, x_n, p_1, \ldots, p_n, t) \tag{4.2}$$

where $j = 1, \ldots, n$ are called the *Hamilton equations* associated with H; they can advantageously be written in compact form as

$$\dot{z} = J\nabla_z H(z, t) \tag{4.3}$$

where ∇_z is the gradient in the $2n$ variables $x_1, \ldots, x_n, p_1, \ldots, p_n$. This observation reduces the study of the existence and uniqueness of the solutions of Hamilton's equations to that of first-order non-autonomous differential systems of ordinary differential equations.

We will assume for simplicity that for every $z_0 = (x_0, p_0)$ belonging to some open subset Ω of $\mathbb{R}^{2n}$ Hamilton's equations have a unique solution $t \longmapsto z(t) = (x(t), p(t))$ such that $z(0) = z_0$, defined for $-T \leq t \leq T$ where $T > 0$.

Example 1 If the Hamiltonian is time-independent of the type

$$H(x, p) = \sum_{j=1}^{n} \frac{p_j^2}{2m_j} + V(x)$$

where the potential V is smooth and satisfies the estimate $V(x) \geq -a|x|^2$ for some $a > 0$ then the solutions of the Hamilton equations exist for all times t and are unique for given initial conditions.

We refer to Abraham and Marsden [1], Chap. 1, Sect. 2.1, for a general discussion of global existence and uniqueness of solutions to Hamilton's equations. When these solutions are defined and unique for some time interval $[-T, T]$ and initial conditions in $\Omega \subset \mathbb{R}^{2n}$ one can define a "partial flow": it is the family $(f_t^H) = (f_t^H)_{-T \leq t \leq T}$ of diffeomorphisms of Ω such that the function $z(t) = f_t^H(z_0)$ is the solution of Hamilton's equations with $z(0) = z_0$. We are actually making here a slight abuse of terminology because $(f_t^H)_{-T \leq t \leq T}$ is, strictly speaking, a true flow (in the sense of the theory of dynamical systems) only when the Hamiltonian H is time-independent. When this is the case, then (f_t^H) is just the phase space flow determined by the Hamiltonian vector field

$$X_H = J\partial_z H = (\partial_x H, -\partial_p H)$$

and we have the group property

$$f_t^H f_{t'}^H = f_{t+t'}^H,\ f_0^H = I \tag{4.4}$$

when t, t' and $t+t'$ are in the interval $[-T, T]$; in particular each f_t^H is a diffeomorphism such that $(f_t^H)^{-1} = f_{-t}^H$. In the general case (i.e. when H is time-dependent), it is better to consider the *time-dependent flow* $(f_{t,t'}^H)$, defined by $f_{t,t'}^H = f_t^H(f_{t'}^H)^{-1}$: it is the mapping which takes the point $(x', p') \in \Omega$ at time t' to the point (x, p) at time t along the solution curve of Hamilton's equation:

$$(x, p) = f_{t,t'}^H(x', p').$$

An essential property of the mappings f_t^H and $f_{t,t'}^H$ is that they are *canonical transformations* (or symplectomorphisms); this means that the Jacobian matrix

$$Df_{t,t'}^H(z') = \frac{\partial(x, p)}{\partial(x', p')} = \begin{pmatrix} \frac{\partial x}{\partial x'} & \frac{\partial x}{\partial p'} \\ \frac{\partial p}{\partial x'} & \frac{\partial p}{\partial p'} \end{pmatrix} \tag{4.5}$$

satisfies the two equivalent conditions

$$Df_{t,t'}^H(z')J(Df_{t,t'}^H(z'))^T = J \tag{4.6}$$

$$Df_{t,t'}^H(z')^T J(Df_{t,t'}^H(z')) = J. \tag{4.7}$$

In other words $Df_{t,t'}^H(z')$ is a symplectic matrix: a real $2n \times 2n$ matrix S is said to be symplectic if it satisfies anyone (and hence both) of the equivalent conditions $SJS^T = J$ or $S^T JS = J$. Symplectic matrices form a group $\mathrm{Sp}(n)$ which will be studied in more detail in Chap. 12 (see de Gosson [7] for an elementary study of Hamiltonian mechanics from the symplectic point of view).

4.2 Free Canonical Transformations

4.2.1 Free Symplectic Matrices

Writing a symplectic matrix in block-form

$$S = \begin{pmatrix} A & B \\ C & D \end{pmatrix}$$

we will say that it is a "free" symplectic matrix if $\det B \neq 0$. Equivalently, the equation $(x, p) = S(x', p')$ can be solved in x and x', which determines uniquely the momenta p and p'. In fact, $(x, p) = S(x', p')$ is equivalent to the linear system

$$x = Ax' + Bp',\ p = Cx' + Dp';$$

solving the first equation in p' yields

$$p' = B^{-1}x - B^{-1}Ax'$$

and inserting this solution is the second equation yields

$$p = DB^{-1}x + (C - DB^{-1}A)x'.$$

This property can be made explicit using the generating function

$$W(x,x') = \tfrac{1}{2}DB^{-1}x^2 - B^{-1}x\cdot x' + \tfrac{1}{2}B^{-1}Ax'^2 \tag{4.8}$$

(see de Gosson [5, 7]). In fact,

$$(x,p) = S(x',p') \Longleftrightarrow \begin{cases} p = \nabla_x W(x,x') \\ p' = -\nabla_{x'} W(x,x') \end{cases} \tag{4.9}$$

as can be checked by a direct calculation. More generally:

Definition 2 We will say that a canonical transformation (symplectomorphism) f defined on a subset Ω of the phase space $\mathbb{R}^{2n}$ is "*free*" if, given x' and x, the equation $(x,p) = f(x',p')$ uniquely determines the momenta p' and p.

The following criterion is a generalization of the condition $\det B \neq 0$ for free symplectic matrices:

Lemma 3 *A canonical transformation f defined on a subset Ω of $\mathbb{R}^{2n}$ is free if and only if the Jacobian matrix*

$$Df(z') = \frac{\partial(x,p)}{\partial(x',p')}$$

is a free symplectic matrix, that is if $\det(\partial x/\partial p') \neq 0$ *on Ω for* $(x,p) = f(x',p')$.

Proof In view of the implicit function theorem the equation $(x,p) = f(x',p')$ can be solved locally in p', for given x' and x, if and only if the Jacobian matrix $\partial x/\partial p'$ is invertible. ■

We refer to the our book [5] for a detailed study, with complete proofs, of the notion of free symplectic matrix. It is proven there, among other properties, that every symplectic matrix can be written as a product of exactly two free symplectic matrices.

4.2.2 Hamilton's Two-Point Characteristic Function

The notion of generating functions extends to arbitrary free canonical transformations, in particular to Hamiltonian flows. We have the following important result, which will allow us to link the notion of action integral to Hamilton–Jacobi theory discussed below:

Proposition 4 *Let M be a real symmetric invertible $n \times n$ matrix, $V = V(x,t)$ a smooth potential function, and set*

$$H(x,p,t) = \sum_{j=1}^{n} \frac{p_j^2}{2m_j} + V(x,t) \tag{4.10}$$

(i) There exists $\varepsilon > 0$ such that $f_{t,t'}^H$ is a free canonical transformation for $0 < |t-t'| < \varepsilon$; (ii) The function $W(x,x',t,t')$ defined by

$$W(x,x',t,t') = \int_\gamma pdx - Hdt \tag{4.11}$$

where the integral is calculated along the phase-space trajectory $s \longmapsto f_{s,t'}^H(x',p')$ joining (x',p') at time t' to (x,p) at time t is a generating function, that is:

$$(x,p) = f_{t,t'}^H(x',p') \Longleftrightarrow \begin{cases} p = \nabla_x W(x,x',t,t') \\ p' = -\nabla_{x'} W(x,x',t,t') \end{cases}.$$

Proof We prove (i) for $n = 1$; the general case is a straightforward generalization ([7], Sect. 4.4.1). The Hamilton equations for H are

$$\dot{x} = p/m \,,\, \dot{p} = -\partial_x V(x,t).$$

Set $(x,p) = f_{t,t'}^H(x',p')$; using a Taylor expansion at time t' we have, taking the Hamilton equations into account,

$$x = x' + m^{-1}p'(t-t') + O((t-t')^2)$$
$$p = p' - \partial_x V(x',t')(t-t') + O((t-t')^2)$$

hence

$$\frac{\partial x}{\partial p'} = m^{-1}(t-t') + O((t-t')^2)$$

which is different from zero for $t - t' \neq 0$ sufficiently small. For a proof of (ii) see de Gosson [7], Sect. 4.4.1, Proposition 92. ■

This result easily extends to the case of Hamiltonians of the type

$$H(x,p,t) = \sum_{j=1}^{n} \frac{1}{2m_j}(p_j - A(x,t))^2 + V(x,t) \tag{4.12}$$

(see de Gosson [7], Sect. 4.4.1).

Definition 5 The function $W(x, x', t, t')$ defined by (4.11) is called *Hamilton's two-point characteristic function,* or a "free generating function" for the flow $(f^H_{t,t'})$.

Here are two examples:

Example 6 Consider the free particle Hamiltonian in one dimension: $H = p^2/2m$. The Hamiltonian flow is defined by

$$(x, p) = f^H_{t,t'}(x', p') = (x' + \tfrac{p'}{m}(t - t'), p') \tag{4.13}$$

hence the integration path in the integral (4.11) is parametrized as $\gamma(s) = (x' + \frac{p'}{m}(s - t'), p')$ and we thus have

$$\int_\gamma pdx - Hdt = \frac{p'^2}{2m}(t - t');$$

making the substitution $p' = m(x - x')/(t - t')$ yields the well-known value

$$W_f(x, x', t, t') = m\frac{(x - x')^2}{2(t - t')} \tag{4.14}$$

for Hamilton's two-point characteristic function for the free particle.

The second example is in a sense more instructive, because it shows that Hamilton's two point function need not be defined for all times:

Example 7 Let H be the one-dimensional harmonic oscillator Hamiltonian:

$$H(x, p) = \frac{1}{2m}(p^2 + m^2\omega^2x^2).$$

Then the generating function exists for all $t - t'$ different from an integer multiple of $2\pi/\omega$ and is given by

$$W(x, x', t, t') = \frac{m\omega}{2\sin\omega(t - t')}\left[(x^2 + x'^2)\cos(\omega(t - t')) - 2xx'\right]. \tag{4.15}$$

The explicit construction of $W(x, x', t, t')$ can be quite complicated outside these elementary cases. For instance, Binder constructs in [3] the two-point characteristic function for the Kepler–Coulomb problem, associated with the Hamiltonian function

$$H = \frac{1}{2m}|\mathbf{p}|^2 - \frac{Ze^2}{|\mathbf{r}|}$$

and this leads to a quite intricate expression.

4.3 The Action of a Dynamical System

The notion of action plays a central role in classical mechanics; it is related to Hamilton's two point characteristic function in a clever way. It can be geometrically understood using the notion of Lagrangian manifold; see de Gosson [7] for a detailed account of this interpretation.

4.3.1 The Poincaré–Cartan Invariant

Let H be an arbitrary (possibly time-independent) Hamiltonian function; the differential one-form

$$\alpha_H = pdx - H(x, p, t)dt \tag{4.16}$$

where $pdx = p_1dx_1 + \cdots + p_ndx_n$ is called the *Poincaré–Cartan integral invariant (or form)*. Note that α_H is a differential form on the extended phase space $\mathbb{R}^n_x \times \mathbb{R}^n_p \times \mathbb{R}_t$ whose restriction to the phase space $\mathbb{R}^n_x \times \mathbb{R}^n_p$ is the canonical 1-form $\beta = pdx$. The exterior derivative $d\beta$ of the latter is just the standard symplectic form $\omega = dp \wedge dx$ (see Arnol'd [2]). The Poincaré–Cartan invariant has the following property, which motivates the terminology "integral invariant": let γ be a loop in extended phase space, and denote by $T(\gamma)$ the two-dimensional tube of trajectories swept out by γ under the action of the Hamiltonian flow $(f^H_{t,t'})$. Let γ' be any loop on $T(\gamma)$ homotopic to γ. Then

$$\oint_\gamma \alpha_H = \oint_{\gamma'} \alpha_H \tag{4.17}$$

(this is an application of the multi-dimensional Stoke's lemma, see Arnol'd [2], Chap. 10 or de Gosson [7], Chap. 5).

The Poincaré–Cartan invariant plays an important role in various areas of Hamiltonian mechanics; we will focus here on its relation with Hamilton–Jacobi's equation, which is a non-linear first-order partial differential equation.

4.3.2 The Action Functional

The notion of action functional is usually defined using the Lagrangian formalism [2, 4, 8]; this approach has certain conceptual advantages and sometimes makes the notion more tractable. We adopt here a slightly different point of view, using directly the Hamiltonian formalism. Let x' and x be two arbitrary points in $\mathbb{R}^n$. We assume that the boundary condition problem

$$\dot{x} = \nabla_p H(x, p, t) \, , \, \dot{p} = -\nabla_x H(x, p, t) \tag{4.18}$$

$$x(t') = x' \, , \, x(t) = x \tag{4.19}$$

has a unique solution $s \longmapsto (x(s), p(s))$; we denote by Γ the lift to the extended phase space $\mathbb{R}^n_x \times \mathbb{R}^n_p \times \mathbb{R}_t$: we have

$$\Gamma(s) = (x(s), p(s), s) \, , t' \leq s \leq t. \tag{4.20}$$

Definition 8 The action functional is by definition the integral

$$S(x, x', t, t') = \int_\gamma pdx - Hdt \tag{4.21}$$

of the Poincaré–Cartan invariant along the "ray" γ, which is the projection on the space-time $\mathbb{R}^n_x \times \mathbb{R}_t$ of the path Γ defined by (4.20).

The idea in the definition above is the following: the point x' being given (at some initial time t'), select among all rays γ departing (with different initial momenta p') from x' at time t' the one which reaches the point x at time t. A little care has to be taken when using the definition above: whereas the generating function $W(x, x', t, t')$ is defined for arbitrary points x, x' (for sufficiently small $|t - t'| > 0$), it can happen that the action functional does not even exist. This is due to the fact that the Hamilton boundary-value problem (4.18)–(4.19) might not be solvable, or even have multi-valued solutions. This difficulty is related to the existence of "central fields of extremals" (see Arnol'd [2], Maslov and Fedoriuk [11]). One can however prove that the action functional always is defined if the points x and x' are close enough. We will make the simplifying assumption that $S(x, x', t, t')$ exists, is single-valued, and infinitely differentiable in all its variables on the subset of $(\mathbb{R}^n_x \times \mathbb{R}_t) \times (\mathbb{R}^n_x \times \mathbb{R}_t)$ where it is defined.

Proposition 9 *For fixed x' and t', set $\phi(x, t) = S(x, x', t, t')$. (i) We have*

$$d\phi(x, t) = pdx - Hdt \tag{4.22}$$

(ii) the function ϕ satisfies the Hamilton–Jacobi equation

$$\frac{\partial \phi}{\partial t} + H(x, \nabla_x \phi) = 0. \tag{4.23}$$

Proof The statement (i) is proven using property (4.17) of the Poincaré–Cartan integral invariant (see Arnol'd [2], Chap. 10, de Gosson [7], Chap. 5). To prove (ii), write

$$d\phi(x, t) = \nabla_x \phi(x, t) dx + \frac{\partial \phi}{\partial t}(x, t) dt;$$

it follows from (4.22) that

$$p = \nabla_x \phi(x,t)\,,\ \frac{\partial \phi}{\partial t}(x,t) = -H(x,p,t)$$

and hence

$$\frac{\partial \phi}{\partial t}(x,t) + H(x, \nabla_x \phi) = \frac{\partial \phi}{\partial t}(x,t) + H(x,p) = 0.$$

■

Here is an illustration:

Example 10 Consider (for $n = 1$) the Hamiltonian $H = \frac{1}{2}p^2x^2$. The Hamilton equations $\dot{x} = px^2$, $\dot{p} = -p^2x$ are easily solved noting that px is a constant of the motion, and one finds

$$x(t) = x(t')e^{p(t')x(t')(t-t')}\,,\ p(t) = p(t')e^{-p(t')x(t')(t-t')}. \tag{4.24}$$

A given point x is reached after time $t - t'$ if one chooses the initial momentum

$$p' = p(t') = \frac{Log(x/x')}{x'(t-t')} \tag{4.25}$$

and the final momentum is

$$p = p(t) = \frac{Log(x/x')}{x(t-t')}.$$

Calculation of the integral (4.21) using the expressions (4.24) of $x(t)$ and $p(t)$ yields

$$\int_\gamma pdx - Hdt = \frac{1}{2}p'^2x'^2. \tag{4.26}$$

Replacing p' with its value (4.25) yields Hamilton's characteristic function:

$$S(x,x',t,t') = \frac{(Log(x/x'))^2}{2(t-t')}; \tag{4.27}$$

it is defined for $t \neq t'$ and $xx' > 0$. One verifies by a straightforward calculation that S satisfies the Hamilton–Jacobi equation (4.23).

We refer to Arnold [2], Calkin [4], Synge and Truesdell [12], for detailed accounts of the Hamilton–Jacobi equation and its usefulness in solving Hamilton's equations. We are more interested here in the relation between this equation and the notion of generating function introduced in the last section.

Proposition 11 *The solution of Hamilton–Jacobi's problem*

$$\frac{\partial \phi}{\partial t} + H(x, \nabla_x \phi) = 0\,,\; \phi(x, t') = \phi'(x) \tag{4.28}$$

is given by

$$\phi(x, t) = \phi'(x') + S(x, x', t, t') \tag{4.29}$$

where the point x' is determined by the condition $(x, p) = f^H_{t,t'}(x', p')$ with $p' = \nabla_x \phi(x', t')$.

Proof See [2], Chap. 10, de Gosson [7], Chap. 5. ∎

4.4 Short-Time Action

While it is usually difficult to find explicit expressions for the action functional, one can obtain in a rather straightforward way short-time approximations. These are, as we will see in the next chapter, very useful in the study of Schrödinger's equation. We begin by showing that the "mid-point rules" frequently used in the theory of the Feynman path integral are not good approximations.

In what follows we will generally use the notation

$$\Delta t = t - t'\,,\; \Delta t^2 = (t - t')^2.$$

We begin by pointing out the inappropriateness of the "mid-point rules" often used in the theory of the Feynman path integral, and show that they lead to incorrect results when used to approximate the action for small times.

4.4.1 On "Mid-Point Rules"

When the Hamiltonian function is of the type "kinetic energy plus potential" it is common practice in the Feynman path integral literature to approximate the generating function S for small values of $t - t'$ by expressions like

$$S_1(x, x', t, t') = m\frac{(x - x')^2}{2\Delta t} - \frac{1}{2}(V(x) + V(x'))\Delta t \tag{4.30}$$

or, alternatively,

$$S_2(x, x', t, t') = m\frac{(x - x')^2}{2\Delta t} - V(\tfrac{1}{2}(x + x'))\Delta t. \tag{4.31}$$

Suppose for instance that H is the harmonic oscillator Hamiltonian considered in Example 7; expanding the terms $\sin\omega(t-t')$ and $\cos\omega(t-t')$ in Taylor series at $t=t'$ yields the *correct* approximation

$$S(x,x',t,t') = m\frac{(x-x')^2}{2\Delta t} - \frac{m\omega^2}{6}(x^2+xx'+x'^2)\Delta t + O(\Delta t^2) \qquad (4.32)$$

to the exact generating function $W=S$ given by formula (4.15). However, if we apply the "rule" (4.30) we get

$$S_1(x,x',t,t') = m\frac{(x-x')^2}{2\Delta t} - \frac{m^2\omega^2}{4}(x^2+x'^2)\Delta t$$

and we thus have

$$S(x,x',t,t') - S_1(x,x',t,t') = O(\Delta t).$$

If we use instead the "rule" (4.31) we get

$$S_2(x,x',t,t') = m\frac{(x-x')^2}{2\Delta t} - \frac{m^2\omega^2}{8}(x+x')^2\Delta t$$

and we have here

$$S(x,x',t,t') - S_2(x,x',t,t') = O(\Delta t).$$

The "mid-point rules" are thus incorrect even to first order, and should therefore be avoided in any rigorous argument.

4.4.2 A Correct Short-Time Approximation

Assume that H is a classical Hamiltonian function of the physical type

$$H(x,p,t) = \sum_{j=1}^{n}\frac{p_j^2}{2m_j} + V(x,t). \qquad (4.33)$$

Proposition 12 *The Hamilton characteristic two-point function for the flow determined by (4.33) is given, for small $\Delta t = t-t' \neq 0$ by the asymptotic expression*

$$S(x,x',t,t') = \sum_{j=1}^{n} m_j\frac{(x_j-x_j')^2}{2\Delta t} - \overline{V}(x,x',t')\Delta t + O(\Delta t^2) \qquad (4.34)$$

where $\overline{V}(x, x', t')$ is the average

$$\overline{V}(x, x', t') = \int_0^1 V(\tau x + (1-\tau)x', t')d\tau. \tag{4.35}$$

Proof See Makri and Miller [9, 10]. In [7], Sect. 4.4.1, we gave an independent proof along the following lines (we limit ourselves to the case $n = 1$): making the Ansatz

$$W = W_{\text{free}} + W_1 \Delta t + W_2 (\Delta t)^2 + \cdots$$

where W_{free} is the generating function of the free particle Hamiltonian (Example 6), one can determine the terms $W_1, W_2, \ldots$ by inserting this expression in the Hamilton–Jacobi equation

$$\frac{\partial W}{\partial t} + \frac{1}{2m}\left(\frac{\partial W}{\partial t}\right)^2 + V = 0.$$

This leads to the equation

$$W_1 + (x - x')\frac{\partial W_1}{\partial x} + V = 0$$

and to similar equations for the lower order terms $W_2, W_3 \ldots$ One next observes that the only smooth solution of the equation above is

$$W_1(x, x', t') = -\frac{1}{x - x'}\int_{x'}^{x} V(x'', t')dx'' = -\overline{V}(x, x', t')$$

where the second equality follows from the first by making the change of variables $x'' = \tau x + (1 - \tau)x'$. ■

Example 13 Consider again the one-dimensional oscillator Hamiltonian

$$H(x, p) = \frac{1}{2m}(p^2 + m^2\omega^2 x^2).$$

We have here $V(x) = \frac{1}{2}m\omega^2 x^2$ and hence

$$\overline{V}(x, x') = \frac{m\omega^2}{6}(x^2 + xx' + x'^2).$$

Using formula (4.34) we recover the approximation (4.32) obtained using Taylor series.

We will write

$$\overline{H}(x,x',p',t')=\sum_{j=1}^{n}\frac{p_j'^2}{2m_j}+\overline{V}(x,x',t')$$

and call the function $\overline{H}$ the "averaged Hamiltonian". We are now going to generalize Proposition 12 to arbitrary Hamiltonian functions.

4.4.3 The Averaged Hamiltonian $\overline{H}$

Let us now consider the case where H is an arbitrary (time-dependent) Hamiltonian function; we do thus not assume it has the form (4.33), or its generalization including a vector potential. For instance, in dimension $n=1$, it could be a monomial $x^r p^s$. We now introduce the following averaged Hamiltonian function:

$$\overline{H}(x,x',p',t')=\int_0^1 H(\tau x+(1-\tau)x',p',t')d\tau. \tag{4.36}$$

Note that when H is of the classical type (4.33) then

$$\overline{H}(x,x',p,t')=\sum_{j=1}^{n}\frac{p_j^2}{2m_j}+\overline{V}(x,x',t) \tag{4.37}$$

where $\overline{V}(x,x',t)$ is defined by formula (4.35) above. Also notice that definition (4.36) can be rewritten as

$$\overline{H}(x,x',p',t')=\frac{1}{t-t'}\int_{t'}^{t} H(x'+v(s-t'),p',t')ds \tag{4.38}$$

where $v=(x-x')/(t-t')$ (make the change of variables $\tau=(s-t')/(t-t')$). This remark makes the physical interpretation of $\overline{H}$ clear: the number $\overline{H}(x,x',p',t')$ is the average value of the energy when a point located at x' at time t' proceeds to the point x at time t with constant velocity v.

The following property of $\overline{H}$ is essential:

Lemma 14 *The function $\overline{H}=\overline{H}(x,x',p',t')$ satisfies, for fixed x' and t', the partial differential equation*

$$\sum_{j=1}^{n}(x_j-x_j')\frac{\partial\overline{H}}{\partial x_j}+\overline{H}=H. \tag{4.39}$$

Proof Performing a partial integration in the integral

$$\overline{H} = \int_0^1 H(\tau x + (1-\tau)x', p', t')d\tau$$

yields

$$\overline{H} = \left[H(\tau x + (1-\tau)x', p', t')\tau\right]_0^1 - \int_0^1 \tau \frac{d}{d\tau} H(\tau x + (1-\tau)x', p', t')d\tau$$

that is

$$\overline{H} = H(x, p, t) - \int_0^1 \tau \frac{d}{d\tau} H(\tau x + (1-\tau)x', p', t')d\tau$$

so that there remains to show that

$$\int_0^1 \tau \frac{d}{d\tau} H(\tau x + (1-\tau)x', p', t')d\tau = \sum_{j=1}^n (x_j - x_j') \frac{\partial \overline{H}}{\partial x_j}. \tag{4.40}$$

Using the chain rule, the integral on the left-hand side is

$$(x_j - x_j') \int_0^1 \tau \frac{\partial H}{\partial x_j}(\tau x + (1-\tau)x', p', t')d\tau = \sum_{j=1}^n (x_j - x_j') \frac{\partial}{\partial x_j} \int_0^1 H(\tau x + (1-\tau)x', p', t')d\tau$$

which proves (4.40). ∎

4.4.4 Short-Time Approximations: The General Case

We now prove the central result of this chapter; it is the key to our discussion of the wavefunction in next chapter:

Proposition 15 *Let* $H \in C^\infty(\mathbb{R}^{2n} \times \mathbb{R})$ *and assume that Hamilton's two-point characteristic function* $S(x, x', t, t')$ *is defined for* $0 < |t - t'| < \varepsilon$ *and* x *and* x' *close enough. Let*

$$\overline{S}(x, x', t, t') = \int_{\overline{\gamma}} pdx - Hdt \tag{4.41}$$

where the path $\overline{\gamma}$ is the projection on space-time of the path

$$\Gamma(s) = (x' + v(s - t'), p', s) \, , \, t' \leq s \leq t$$

where $v = (x - x')/(t - t')$ is the average velocity and p' is determined by (4.18)–(4.19).

(i) The function $\overline{S}$ has the second order expansion

$$\overline{S}(x, x', t, t') = p'(x - x') - \overline{H}(x, x', p', t')\Delta t + O(\Delta t^2) \tag{4.42}$$

where $\overline{H}$ is defined by (4.36) and $\Delta t = t - t'$; when H is time-independent the term $O(\Delta t^2)$ cancels.

(ii) We have

$$S(x, x', t, t') = \overline{S}(x, x', t, t') + O(\Delta t^2). \tag{4.43}$$

Proof (i) Let us prove formula (4.42). We have, by definition of v,

$$\int_{\overline{\gamma}} pdx = \int_{t'}^{t} p'vds = p'(x - x'). \tag{4.44}$$

On the other hand, by definition of the integration path $\overline{\gamma}$,

$$\int_{\overline{\gamma}} Hdt = \int_{t'}^{t} H(x' + v(s - t'), p', s)ds.$$

A first order Taylor expansion at $s = t'$ yields

$$H(x' + v(s - t'), p', s) = H(x' + v(s - t'), p', t') + O(s - t')$$

and hence

$$\begin{aligned}\int_{t'}^{t} H(x' + v(s - t'), p', s)ds &= \int_{t'}^{t} H(x' + v(s - t'), p', t')ds + O(\Delta t^2) \\ &= \overline{H}(x, x', p', t')(t - t') + O(\Delta t^2).\end{aligned}$$

It follows that

$$\int_{\overline{\gamma}} Hdt = \int_{t'}^{t} H(x' + v(s - t'), p', t')ds + O(\Delta t^2)$$
$$= \overline{H}(x, x', p', t')(t - t') + O(\Delta t^2)$$

hence formula (4.42) taking (4.44) into account. (ii) Let us next prove the asymptotic formula (4.43). Writing

$$p(s) = p' + (s - t')\dot{p}(t') + O((s - t')^2)$$

we have, the path γ being defined as in (4.21),

$$\int_{\gamma} pdx = \int_{t'}^{t} p'\dot{x}(s)ds + O(\Delta t^2) = p'(x - x') + O(\Delta t^2)$$

hence, taking (4.44) into account,

$$\int_{\gamma} pdx - \int_{\overline{\gamma}} pdx = O(\Delta t^2). \tag{4.45}$$

Let us next evaluate the difference

$$\Delta(\gamma, \overline{\gamma}) = \int_{\gamma} Hdt - \int_{\overline{\gamma}} Hdt;$$

we will show that $\Delta(\gamma, \overline{\gamma}) = O(\Delta t^2)$, which will prove (4.43). We have

$$\Delta(\gamma, \overline{\gamma}) = \int_{t'}^{t} H(x(s), p(s), s)ds - \int_{t'}^{t} H(x' + v(s - t'), p', t')ds;$$

since

$$H(x(s), p(s), s) = H(x(s), p(s), t') + O(s - t')$$

we get

$$\Delta(\gamma, \overline{\gamma}) = \int_{t'}^{t} H(x(s), p(s), t')ds$$
$$- \int_{t'}^{t} H(x' + v(s - t'), p', t')ds + O(\Delta t^2).$$

In view of formula (4.39) in Lemma 14 above, we have

$$\int_{t'}^{t} H(x(s), p(s), t')ds = \sum_{j=1}^{n} \int_{t'}^{t} (x_j(s) - x'_j)\frac{\partial \overline{H}}{\partial x_j}(x(s), x', p', t')ds$$
$$+ \int_{t'}^{t} \overline{H}(x(s), x', p', t')ds;$$

since $x_j(s) - x'_j = O((s - t'))^2$ and the partial derivatives $\partial \overline{H}/\partial x_j$ are bounded for s in the interval $[t', t]$ we have

$$\int_{t'}^{t} (x_j(s) - x'_j)\frac{\partial \overline{H}}{\partial x_j}(x(s), x', p', t')ds = O(\Delta t^2)$$

for $j = 1, \ldots, n$, and hence

$$\int_{t'}^{t} H(x(s), p(s), t')ds = \int_{t'}^{t} \overline{H}(x(s), x', p', t')ds + O(\Delta t^2). \qquad (4.46)$$

Writing

$$x(s) = x' + \dot{x}(t')(s - t) + O((s - t')^2)$$

and observing that

$$v = \frac{x - x'}{t - t'} = \dot{x}(t') + O((t - t')^2)$$

we have

$$x(s) = x' + v(s - t) + O((s - t')^2).$$

It follows that

$$\overline{H}(x(s), x', p', t') = \overline{H}(x' + v(s - t), p', t') + O((s - t')^2)$$

and hence, integrating this expansion from t' to t,

$$\int_{t'}^{t} \overline{H}(x(s), x', p', t')ds = \int_{t'}^{t} \overline{H}(x' + v(s - t), p', t')ds + O(\Delta t^2)$$

so that $\Delta(\gamma, \overline{\gamma}) = O(\Delta t^2)$, as we set out to prove. ∎

For Hamiltonian functions of the classical type

$$H(x,p,t)=\sum_{j=1}^{n}\frac{p_j^2}{2m_j}+V(x,t) \tag{4.47}$$

formula (4.42) reduces to the approximation (4.34). Let us check this for $n=1$. We have

$$\overline{H}(x,x',t,t')=\frac{p^2}{2m}+\overline{V}(x,x',t,t')$$

where $\overline{V}$ is defined by the integral (4.35). There remains to check that

$$p(x-x')-\frac{p^2}{2m}=m\frac{(x-x')^2}{2\Delta t}+O(\Delta t^2) \tag{4.48}$$

where p is defined as follows: for $\Delta t=t-t'$ sufficiently, the point x is reached at time t by the projection of a unique Hamiltonian trajectory $s\longmapsto(x(s),p(s))$ $(t'\leq s\leq t)$ starting from the point x' at time t', and this trajectory fixes once for all the initial and final momenta p' and p. Writing $x=x(t)$, $x'=x(t')$, $\dot{x}'=\dot{x}(t')$ we have

$$x=x'+\dot{x}'\Delta t+O(\Delta t^2)$$

and, in view of Hamilton's equation for the position coordinate,

$$\dot{x}'=\frac{1}{m}p'=\frac{1}{m}p+O(\Delta t)$$

and hence

$$x-x'=\frac{1}{m}p+O(\Delta t^2);$$

solving this equation in p and inserting the found value in $p(x-x')-p^2/2m$ yields the approximation (4.48).

Example 16 Let us return to the monomial $H=\frac{1}{2}p^2x^2$ considered in Example 10. We have

$$\overline{H}(x,x',p',t')=\frac{1}{6}p'^2(x^2+xx'+x'^2).$$

Formula (4.42) yields, since H is time-independent,

$$\overline{S}(x,x',t,t')=p'(x-x')-\frac{1}{6}p'^2(x^2+xx'+x'^2)\Delta t.$$

Using the first formula (4.24) we have

$$x = x'(1 + p'x'\Delta t) + O(\Delta t^2)$$

and hence

$$\overline{S}(x, x', t, t') = \frac{1}{2}p'^2 x'^2 \Delta t.$$

Inserting the value (4.25) of the initial momentum p' yields

$$\overline{S}(x, x', t, t') = \frac{(Log(x/x'))^2}{2\Delta t} + O(\Delta t^2)$$

that is $\overline{S} = S + O(\Delta t^2)$ as predicted by formula (4.43) in Proposition 15 (in fact we have $\overline{S} = S$ here).

We are going to use these results to construct short-time approximations to the wavefunction.

References

1. R. Abraham, J.E. Marsden, *Foundations of Mechanics*, 2nd edn. (The Benjamin/Cummings Publishing Company, 1978)
2. V.I. Arnold, *Mathematical Methods of Classical Mechanics, Graduate Texts in Mathematics*, 2nd edn. (Springer, New York, 1989)
3. S.M. Binder, Two-point characteristic function for the Kepler-Coulomb problem. J. Math. Phys. **16**(10), 2000–2004 (1975)
4. M.G. Calkin, R. Weinstock, *Lagrangian and Hamiltonian Mechanics* (World Scientific, Singapore, 1996)
5. M. de Gosson, *Symplectic Geometry and Quantum Mechanics*. Operator Theory: Advances and Applications. Advances in Partial Differential Equations, vol. 166 (Birkhäuser, Basel, 2006)
6. M. de Gosson, *Symplectic Methods in Harmonic Analysis and in Mathematical Physics* (Birkhäuser, 2011)
7. M. de Gosson, *The Principles of Newtonian and Quantum Mechanics: The Need for Planck's Constant, h (With a foreword by B. Hiley)* (Imperial College Press, World Scientific, 2001)
8. H. Goldstein, *Classical Mechanics* (Addison–Wesley, 1950); 2nd edn. (1980); 3rd edn. (2002)
9. N. Makri, Feynman path integration in quantum dynamics. Comput. Phys. Comm. **63**(1), 389–414 (1991)
10. N. Makri, W.H. Miller, Correct short time propagator for Feynman path integration by power series expansion in Δt. Chem. Phys. Lett. **151**, 1–8 (1988)
11. V.P. Maslov, M.V. Fedoriuk, Semi-classical approximation in quantum mechanics, vol. 7. (Springer Science & Business Media, 2001) (D. Reidel, Boston, 1981)
12. J.L. Synge, C. Truesdell, *Handbuch der Physik: Prinzipien der Klassischen Mechanik und Feldtheorie*, vol. 3 (Springer, 1960)

Using the formula (4.24) we have

$$[illegible]$$

and hence

$$[illegible]$$

Therefore [illegible] (4.4) [illegible]

$$[illegible]$$

[illegible]

[illegible]

References

1. [illegible]
2. [illegible]
3. [illegible]
4. [illegible]
5. [illegible]
6. [illegible]
7. [illegible]
8. [illegible]
9. [illegible]
10. [illegible]
11. [illegible]
12. [illegible]

Chapter 5
Wave Mechanics and the Schrödinger Equation

While we have been concerned in Chap. 3 with the quantization of monomials, we now address the question of how we should quantize an arbitrary dynamical variable (i.e. a Hamiltonian function). While Born and Jordan's original argument was purely algebraic, and strictly speaking limited to the case of polynomials, there is another way to justify (both mathematically and physically) their quantization scheme. This will be done by viewing the notion of wavefunction as the fundamental object, and by imposing a natural physical condition on its short-time evolution. This will lead us to the Schrödinger equation, provided that we use the Born–Jordan scheme in quantizing the Hamiltonian function. This chapter thus gives a new justification for the appropriateness of the use of Born–Jordan quantization. The main formula expressing the quantification of the Hamiltonian is actually the starting point to the pseudo-differential approach to Born–Jordan quantization as will be developed in Chap. 10.

5.1 Matter Waves

A quantum system is described by a quantum state $|\psi\rangle$; its position representation $\langle x|\psi\rangle \equiv \psi(x)$ is the *wavefunction* in configuration space $\mathbb{R}^n_x$. We will follow here Heisenberg's point of view, in which the wavefunction is viewed as a collection of potentialities (Peres [5], Shimony [6]).

5.1.1 The Free Particle

Consider a particle with mass m moving freely with velocity $\mathbf{v} = (v_x, v_y, v_z)$ in configuration space $\mathbb{R}^3$. Following de Broglie's matter wave postulate, to this particle is associated a plane wave with phase

$$\Theta(\mathbf{r}, t) = \mathbf{k} \cdot \mathbf{r} - \omega t + C.$$

M.A. de Gosson, *Born–Jordan Quantization*, Fundamental Theories of Physics 182, DOI 10.1007/978-3-319-27902-2_5

Here C is an arbitrary constant, and the wave vector $\mathbf{k}$ and the frequency $\omega(\mathbf{k})$ are defined by the relativistic equations

$$\mathbf{k} = m\mathbf{v}/\hbar\,,\ \omega = mc^2/\hbar$$

so that we have, writing $\mathbf{r} = (x, y, z)$,

$$\Theta(\mathbf{r}, t) = \frac{1}{\hbar}\left(\mathbf{p}\cdot\mathbf{r} - mc^2 t\right) + C$$

where $\mathbf{p} = m\mathbf{v} = \hbar\mathbf{k}$ is the momentum vector. For small velocities the energy

$$mc^2 = \frac{m_0 c^2}{\sqrt{1-(v/c)^2}}$$

is approximated by $m_0 c^2 + \frac{1}{2} m_0 v^2$ with $v = |\mathbf{v}|$ hence we can write, neglecting terms $O(v^4/c^2)$,

$$\Theta(\mathbf{r}, t) = \frac{1}{\hbar}\phi(\mathbf{r}, t) - \frac{m_0 c^2}{\hbar} t$$

where ϕ is the action function:

$$\phi(\mathbf{r}, t) = \mathbf{p}_0 \cdot \mathbf{r} - \frac{p_0^2}{2m} t + C\hbar$$

($\mathbf{p}_0 = m_0\mathbf{v}$ and $p_0 = |\mathbf{p}_0|$). Now, there is no point in keeping the term $m_0 c^2 t/\hbar$ (its presence affects neither the phase nor the group velocities) so we can take as definition of the phase of the matter wave the function

$$\Theta(\mathbf{r}, t) = \frac{1}{\hbar}\left(\phi(\mathbf{r}, t) + C\right)$$

and fix the constant C by requiring that, at initial time $t = 0$, the equation $\Theta = 0$ determines the phase plane $\mathbf{p}\cdot\mathbf{r} = \mathbf{p}_0 \cdot \mathbf{r}_0$. This leads to the expression $\phi(\mathbf{r}, t) = S_{\mathbf{r}_0,\mathbf{p}_0}(\mathbf{r}, t)$ where

$$S_{\mathbf{r}_0,\mathbf{p}_0}(\mathbf{r}, t) = \mathbf{p}_0 \cdot (\mathbf{r} - \mathbf{r}_0) - \frac{p'^2}{2m} t \tag{5.1}$$

showing that ϕ is the gain in action when the particle proceeds from $\mathbf{r}_0$ at time $t = 0$ to $\mathbf{r}$ at time t with constant velocity $\mathbf{v}_0 = \mathbf{p}_0/m$ (we are writing $m = m_0$). We next observe that the momentum vector $\mathbf{p}_0$ is quite arbitrary, and as long as it has not been measured, all the potentialities associated with the phase (5.1) are present. We therefore define the wavefunction of the free particle by summing over all these potentialities:

$$\psi_{\mathbf{r}_0}(\mathbf{r},t) = \left(\tfrac{1}{2\pi\hbar}\right)^3 \iiint e^{\frac{i}{\hbar} S_{\mathbf{r}_0,\mathbf{p}_0}(\mathbf{r},t)} d^3\mathbf{p}_0 \tag{5.2}$$

where $d^3\mathbf{p} = dp_x dp_y dp_z$ (we will see the reason for the normalizing prefactor in a moment).

Observing that the function $S_{\mathbf{r}_0,\mathbf{p}_0}(\mathbf{r},t)$ is a solution of the Hamilton–Jacobi equation for the free particle Hamiltonian

$$\frac{\partial}{\partial t} S_{\mathbf{r}_0,\mathbf{p}_0} + \frac{1}{2m}\left(\nabla_{\mathbf{r}} S_{\mathbf{r}_0,\mathbf{p}_0}\right)^2 = 0 \tag{5.3}$$

with initial condition

$$S_{\mathbf{r}_0,\mathbf{p}_0}(\mathbf{r},0) = \mathbf{p}_0 \cdot (\mathbf{r} - \mathbf{r}_0) \tag{5.4}$$

one immediately verifies that, for fixed $\mathbf{r}'$ and t' the wavefunction $\psi_{\mathbf{r}_0}(\mathbf{r},t)$ satisfies Schrödinger's equation

$$i\hbar \frac{\partial}{\partial t}\psi_{\mathbf{r}_0} = -\frac{\hbar^2}{2m}\nabla^2_{\mathbf{r}}\psi_{\mathbf{r}_0}; \tag{5.5}$$

moreover

$$\lim_{t\to 0}\psi_{\mathbf{r}_0}(\mathbf{r},t) = \left(\tfrac{1}{2\pi\hbar}\right)^3 \iiint e^{\frac{i}{\hbar}\mathbf{p}_0\cdot(\mathbf{r}-\mathbf{r}_0)} d^3\mathbf{p}_0$$

hence, by the properties of the Fourier transform

$$\psi_{\mathbf{r}',t'}(\mathbf{r},0) = \delta(\mathbf{r} - \mathbf{r}_0).$$

The wavefunction $\psi_{\mathbf{r}_0}$ thus represents a wave emanating at time $t = 0$ from a point-like source placed at $\mathbf{r}_0 = (x_0, y_0, z_0)$.

The construction above extends *mutatis mutandis* to the case of an arbitrary number of degrees of freedom, and with arbitrary time origin t'. Introducing generalized coordinates $x = (x_1, ..., x_n)$ and $p = (p_1, ..., n_n)$ one replaces the phase (5.1) with

$$S(x, x', p', t, t') = p'(x - x') - \sum_{j=1}^{n} \frac{p_j'^2}{2m_j}(t - t'); \tag{5.6}$$

this function satisfies the Hamilton–Jacobi equation

$$\frac{\partial S}{\partial t} + \sum_{j=1}^{n} \frac{1}{2m_j}\left(\frac{\partial S}{\partial x_j}\right)^2 = 0\,,\ S_{|t=t'} = p'(x - x').$$

It follows that the function

$$\psi_{x',t'}(x,t) = \left(\tfrac{1}{2\pi\hbar}\right)^n \int e^{\frac{i}{\hbar}S(x,x',p',t,t')} d^n p' \tag{5.7}$$

satisfies the Schrödinger equation

$$i\hbar \frac{\partial \psi_{x',t'}}{\partial t} = -\sum_{j=1}^{n} \frac{\hbar^2}{2m_j} \frac{\partial^2 \psi_{x',t'}}{\partial x_j^2} \tag{5.8}$$

with initial datum

$$\psi_{x',t'}(x,t') = \delta(x - x').$$

5.1.2 *The Free Propagator*

We now introduce the following notation: we write

$$K_{\mathrm{f}}(x,x',t,t') = \psi_{x',t'}(x,t)$$

(the subscript f standing for "free") and call the function K_{f} the *free particle propagator*; in Dirac bra-ket notation

$$K_{\mathrm{f}}(x,x',t,t') = \langle x,t|\widehat{H}|x',t'\rangle$$

is thus a "matrix element". The knowledge of K_{f} allows us to solve the Schrödinger equation

$$i\hbar \frac{\partial \psi}{\partial t} = -\sum_{j=1}^{n} \frac{\hbar^2}{2m_j} \frac{\partial^2 \psi}{\partial x_j^2} \tag{5.9}$$

for arbitrary initial condition ψ' at time t'; in fact:

$$\psi(x,t) = \int K_{\mathrm{f}}(x,x',t,t')\psi'(x')d^n x';$$

therefore K_{f} is also often called, especially in the mathematical literature, a *distributional kernel.* Notice that the condition $\psi(x,t') = \psi'(x)$ is equivalent to the condition

$$\lim_{t\to t'} K_{\mathrm{f}}(x,x',t,t') = \delta(x - x'). \tag{5.10}$$

We next note that the integral in the right-hand side of (5.7) can be easily evaluated using the theory of Fresnel integrals, and one finds the explicit expression

$$K_{\mathrm{f}}(x,x',t,t') = \left(\frac{M}{2\pi i\hbar(t-t')}\right)^{n/2} e^{\frac{i}{\hbar}W_{\mathrm{f}}(x,x',t,t')} \tag{5.11}$$

where $M = m_1 \cdots m_n$ and

$$W_{\mathrm{f}}(x,x',t,t') = \sum_{j=1}^{n} m_j \frac{(x_j - x'_j)^2}{2(t-t')}; \tag{5.12}$$

one immediately recognizes Hamilton's two-point characteristic function (4.14) for the free particle. The prefactor in the right-hand side of formula (5.11) can be written

$$\left(\frac{M}{2\pi i\hbar(t-t')}\right)^{n/2} = \left(\frac{1}{2\pi i\hbar}\right)^{n/2} \sqrt{\det(-W''_{\mathrm{f}})} \tag{5.13}$$

where $\det(-W''_{\mathrm{f}})$ is the Van Vleck determinant, i.e. the determinant of the Hessian (= matrix of second mixed derivatives) of W_{f}; in both formulas (5.11) and (5.13) the argument of the square roots are chosen so that the initial condition (5.10) holds; this is achieved by defining $\arg i = \pi/2$ and

$$\arg(t-t') = \begin{cases} 0 & \text{if } t-t' > 0 \\ \pi & \text{if } t-t' < 0 \end{cases} \tag{5.14}$$

(see de Gosson [1, 2] for a detailed study of the "right choice" of argument, which is closely related to the theory of the Maslov index). With this choice one has the explicit expression

$$\left(\frac{M}{2\pi i\hbar(t-t')}\right)^{n/2} = e^{-in\pi/4} i^{m(t-t')} \left(\frac{M}{2\pi\hbar|t-t'|}\right)^{n/2} \tag{5.15}$$

the "Maslov index" $m(t-t')$ being defined by

$$m(t-t') = \begin{cases} 0 & \text{if } t-t' > 0 \\ 2n & \text{if } t-t' < 0 \end{cases}. \tag{5.16}$$

5.2 The General Case

5.2.1 *The Abstract Schrödinger Equation*

For simplicity we limit ourselves here to time-independent Hamiltonian functions; the results will be generalized later to the time-dependent case as well. We assume that the time evolution of an initial wavefunction $\psi'(x) = \psi(x,t')$ is governed by a strongly continuous one-parameter group (U_t) of unitary operators acting on $L^2(\mathbb{R}^n)$:

$$\psi(x,t) = U_{t-t'}\psi'(x).$$

It follows from Stone's theorem [7] on unitary operators that there exists a (generally unbounded) self-adjoint operator $\widehat{H}$ such that $U_t = e^{-i\widehat{H}t/\hbar}$; equivalently the one-parameter group (U_t) satisfies the operator equation

$$i\hbar\frac{d}{dt}U_t = \widehat{H}U_t. \tag{5.17}$$

As a consequence, the wavefunction ψ satisfies the Schrödinger equation

$$i\hbar\frac{\partial\psi}{\partial t} = \widehat{H}\psi,\ \psi(x,t') = \psi'(x) \tag{5.18}$$

where the operator $\widehat{H}$ is sofar unknown. The next step consists in remarking that since the Schwartz space $\mathcal{S}(\mathbb{R}^n)$ is a (dense) subspace of $L^2(\mathbb{R}^n)$ and $L^2(\mathbb{R}^n)$ is a subspace of $\mathcal{S}'(\mathbb{R}^n)$, the operators U_t are de facto continuous operators $\mathcal{S}(\mathbb{R}^n) \longrightarrow \mathcal{S}'(\mathbb{R}^n)$. It follows (Schwartz's kernel theorem, see e.g. Hörmander [3]) that there exists a distribution $K = K(x,x',t,t')$ such that

$$\psi(x,t) = \int K(x,x',t,t')\psi'(x')d^nx'; \tag{5.19}$$

in Dirac's notation $K(x,x',t,t')$ is just the matrix element $\langle x,t|\widehat{H}|x',t'\rangle$. Since $\psi(x,t') = \psi'(x')$ we must have, in addition,

$$\lim_{t\to t'} K(x,x',t,t') = \delta(x-x'). \tag{5.20}$$

We will view, as in the free particle case, the propagator K as the wavefunction $\psi_{x',t'}$ of a quantum system located at the point x' at initial time t':

$$K(x,x',t,t') = \psi_{x',t'}(x,t).$$

We emphasize that we have so far rigorously justified the existence of a "Schrödinger equation" satisfied by the matter wave, and that formula (5.19) proving the existence of a propagator is a purely *mathematical* consequence of Stone's theorem and of Schwartz's kernel theorem. But there is no clue to what the operator $\widehat{H}$ should be; for this we will need to make a physical assumption.

5.2.2 The Approximate Wavefunction

Let us come back to the case of a point-like source located at a point x' in n-dimensional configuration space $\mathbb{R}^n_x$. We now assume that the classical motion is governed by an arbitrary Hamiltonian function; it can be a function of the classical type

$$H(x,p,t)=\sum_{j=1}^{n}\frac{p_j^2}{2m_j}+V(x,t)$$

where the potential is smooth function in the variables x and t, but it can also be a totally arbitrary function of the variables x, p, t: we do not suppose that H is of any particular " physical" type.

We would like to determine the corresponding wavefunction $\psi_{x',t'}$. A first guess is that the phase of the wavefunction should be the "obvious" generalization

$$S(x,x',p',t,t')=p'(x-x')-H(x,p',t')(t-t')$$

of (5.6), in which case we could take as propagator the analogue

$$\psi_{x',t'}(x,t)=\left(\tfrac{1}{2\pi\hbar}\right)^n\int e^{\frac{i}{\hbar}S(x,x',p',t,t')}d^np'$$

of (5.7). However, this is not a very good guess; it is easy to see that such a wavefunction does not satisfy a linear evolution equation (see for instance the analysis in de Gosson [1], especially Chap. 8). What we shall do instead, is to *postulate* that the correct phase S is, for small time intervals $t-t'$, given by the expression

$$S(x,x',p',t,t')=\overline{S}(x,x',p',t,t')+O((t-t')^2) \tag{5.21}$$

where

$$\overline{S}(x,x',p',t,t')=p'(x-x')-\overline{H}(x,x',p',t')(t-t') \tag{5.22}$$

the function $\overline{H}$ being given by formula (4.36), that is

$$\overline{H}(x,x',p,t')=\int_0^1 H(\tau x+(1-\tau)x',p,t')d\tau. \tag{5.23}$$

We will see in a moment that this postulate will allow us to determine the Schrödinger equation, but let us first discuss the validity of (5.21) on a simple example which extends the case of the free particle. Let H be the harmonic oscillator Hamiltonian in one dimension:

$$H(x,p)=\frac{p^2}{2m}+\frac{1}{2}m\omega^2x^2.$$

The corresponding exact propagator is well-known: choosing $t' = 0$ for simplicity, it is

$$K_{\mathrm{ho}}(x, x', t) = \sqrt{\frac{m\omega}{2\pi i \hbar \sin \omega t}} e^{\frac{i}{\hbar} W(x,x',t)} \tag{5.24}$$

where $W(x, x', t)$ is given by formula (4.15):

$$W(x, x', t) = \frac{m\omega}{2 \sin \omega t} \left[(x^2 + x'^2) \cos \omega t - 2xx'\right] \tag{5.25}$$

(it is essentially the Mehler formula [4, 8], and can be directly obtained by using the metaplectic representation of the symplectic group we discuss in Chap. 12 (also see de Gosson [1, 2]). The argument of the square root is chosen so that K_{ho} reduces in the limit $t \to 0$ to $\delta(x - x')$. For small values of t we have

$$\sqrt{\frac{m\omega}{2\pi i \hbar \sin \omega t}} = \sqrt{\frac{m}{2\pi i \hbar t}} + O(t^2)$$

and $W(x, x', t) = W(x, x', t, 0)$ is given by $W = \overline{W} + O(t^2)$ where

$$\overline{W}(x, x', t) = m\frac{(x - x')^2}{2t} - \frac{m\omega^2}{6}(x^2 + xx' + x'^2)t$$

(formula (4.32) in Chap. 4). Combining these two formulas, we have $K_{\mathrm{ho}} = \overline{K}_{\mathrm{ho}} + O(t^2)$ where

$$\overline{K}_{\mathrm{ho}}(x, x', t) = \sqrt{\frac{m}{2\pi i \hbar t}} \times \exp\left[\frac{i}{\hbar}\left(m\frac{(x - x')^2}{2t} - \frac{m\omega^2}{6}(x^2 + xx' + x'^2)t\right)\right];$$

using the Fresnel formula, as in the study of the free particle, we have

$$\sqrt{\frac{m}{2\pi i \hbar t}} \exp\left[\frac{i}{\hbar} m \frac{(x-x')^2}{2t}\right] = \frac{1}{2\pi \hbar} \int e^{\frac{i}{\hbar} p'(x-x')} e^{\frac{i}{2m\hbar} p'^2 (x-x')} dp'$$

and hence

$$\overline{K}_{\mathrm{ho}}(x, x', t) = \left(\frac{1}{2\pi \hbar}\right)^n \int e^{\frac{i}{\hbar}(p'(x-x') - \overline{H}(x,x',p',t))} dp'$$

where

$$\overline{H}(x, x', p', t) = \frac{p^2}{2m} - \frac{m\omega^2}{6}(x^2 + xx' + x'^2)t.$$

It follows, as in the case of the free propagator, that for small values of time t, the harmonic oscillator propagator is obtained by integrating over all possible momenta for a wave with phase (5.22). We leave it to the reader, as a pleasant exercise, to verify that the approximation formulas above cannot be obtained by using any of the "mid-point rules" briefly discussed in Sect. 4.4.1.

5.2.3 Back to Schrödinger's Equation

The postulate above amounts to assuming that the exact wavefunction $\psi_{x',t'}$ is approximated, for small time intervals $t - t'$, by the function

$$\overline{\psi}_{x',t'}(x,t) = \left(\tfrac{1}{2\pi\hbar}\right)^n \int e^{\frac{i}{\hbar}\overline{S}(x,x',p',t,t')} d^n p'; \tag{5.26}$$

since the true (but also unknown) phase S is approximated by $\overline{S}$ to order $O((t-t')^2)$ we have

$$\psi_{x',t'}(x,t) = \overline{\psi}_{x',t'}(x,t) + O((t-t')^2). \tag{5.27}$$

Observe that this is tantamount to assuming that the exact propagator K for the abstract Schrödinger equation

$$i\hbar\frac{\partial\psi}{\partial t} = \widehat{H}\psi \, , \, \psi(x,t') = \psi'(x). \tag{5.28}$$

is approximated by

$$\overline{K}(x,x',t,t') = \left(\tfrac{1}{2\pi\hbar}\right)^n \int e^{\frac{i}{\hbar}\overline{S}(x,x',p',t,t')} d^n p'; \tag{5.29}$$

in fact

$$K(x,x',t,t') = \overline{K}(x,x',t,t') + O((t-t')^2). \tag{5.30}$$

We claim that this is enough to determine unambiguously the Hamiltonian operator $\widehat{H}$. The following result is essential, because it justifies and extends, as we will see, Born–Jordan quantization:

Proposition 1 *Suppose that the propagator K satisfies the asymptotic formula (5.30) where $\overline{K}$ is given by (5.29). Then the operator $\widehat{H}$ is given by*

$$\widehat{H}\psi(x) = \left(\tfrac{1}{2\pi\hbar}\right)^n \int e^{\frac{i}{\hbar}p'(x-x')}\overline{H}(x,x',p',t')\psi(x')d^n p' d^n x' \tag{5.31}$$

where $\overline{H}$ is given by formula (4.36):

$$\overline{H}(x, x', p', t') = \int_0^1 H(\tau x + (1 - \tau)x', p', t')d\tau; \tag{5.32}$$

Proof (i) We begin by observing that the knowledge of the propagator K determines the operator $\widehat{H}$: we have

$$\psi(x, t) = \int K(x, x', t, t')\psi(x, t')d^n x'$$

and hence

$$i\hbar \frac{\partial \psi}{\partial t}(x, t) = i\hbar \int \frac{\partial K}{\partial t}(x, x', t, t')\psi(x, t')d^n x';$$

taking the abstract Schrödinger equation (5.18) into account, we get

$$\widehat{H}\psi(x, t) = i\hbar \int \frac{\partial K}{\partial t}(x, x', t, t')\psi(x, t')d^n x'.$$

Using (5.30) it follows that we have

$$\widehat{H}\psi(x, t) = i\hbar \int \frac{\partial \overline{K}}{\partial t}(x, x', t, t')\psi(x, t')d^n x' + O(t - t')$$

and hence, letting $t \to t'$,

$$\widehat{H}\psi(x, t') = i\hbar \int \frac{\partial \overline{K}}{\partial t}(x, x', t, t)\psi(x, t')d^n x'.$$

Now, by definition (5.29) of $\overline{K}$,

$$\frac{\partial \overline{K}}{\partial t}(x, x', t, t') = \left(\tfrac{1}{2\pi\hbar}\right)^n \tfrac{i}{\hbar} \int e^{\frac{i}{\hbar}\overline{S}(x,x',t,t')} \frac{\partial \overline{S}}{\partial t}(x, x', t, t')d^n p'$$

Taking into account the definition (5.22) of the approximate phase $\overline{S}$, we have,

$$\frac{\partial \overline{S}}{\partial t}(x, x', t, t') = -\overline{H}(x, x', p', t')$$

so that

$$\frac{\partial \overline{K}}{\partial t}(x, x', t, t') = \left(\tfrac{1}{2\pi\hbar}\right)^n \tfrac{1}{i\hbar} \int e^{\frac{i}{\hbar}\overline{S}(x,x',t,t')}\overline{H}(x, x', p', t')d^n p'.$$

Taking the limit $t \to t'$ and multiplying both sides by $i\hbar$ we finally get

$$\lim_{t \to t'} \left[i\hbar \frac{\partial \overline{K}}{\partial t}(x, x', t, t') \right] = \left(\tfrac{1}{2\pi\hbar}\right)^n \int e^{\frac{i}{\hbar}p(x-x')} \overline{H}(x, x', p', t') d^n p'$$

which proves (5.31). □

When the Hamiltonian function is of the usual type "kinetic energy plus potential" the result above allows us to recover the usual text-book version of Schrödinger's equation:

Proposition 2 *When the Hamiltonian function is of the classical type*

$$H(x, p, t) = \sum_{j=1}^{n} \frac{p_j^2}{2m_j} + V(x, t) \tag{5.33}$$

where V is a smooth potential, then the operator $\widehat{H}$ is given by

$$\widehat{H} = \sum_{j=1}^{n} \frac{-\hbar^2}{2m_j} \frac{\partial^2}{\partial x_j^2} + V(x, t). \tag{5.34}$$

Proof We have here

$$\overline{H}(x, x', p', t') = \sum_{j=1}^{n} \frac{p_j^2}{2m_j} + \overline{V}(x, x', t')$$

where $\overline{V}(x, x', t')$ is the averaged potential (4.35):

$$\overline{V}(x, x', t') = \int_0^1 V(\tau x + (1 - \tau)x', t') d\tau.$$

Formula (5.31) becomes, using the properties of the Fourier transform,

$$\widehat{H}\psi(x) = \sum_{j=1}^{n} \frac{-\hbar^2}{2m_j} \frac{\partial^2 \psi}{\partial x_j^2}(x)$$
$$+ \left(\tfrac{1}{2\pi\hbar}\right)^n \int e^{\frac{i}{\hbar}p'(x-x')} \overline{V}(x, x', t') \psi(x') d^n p' d^n x'.$$

Since $\overline{V}(x, x, t') = V(x, t')$ we have,

$$\begin{aligned}\int e^{\frac{i}{\hbar}p'(x-x')}\overline{V}(x,x',t')\psi(x')d^np'd^nx' &= \int \overline{V}(x,x',t')\psi(x')\left(\int e^{\frac{i}{\hbar}p'(x-x')}d^np'\right)d^nx' \\ &= (2\pi\hbar)^n\int \overline{V}(x,x',t')\psi(x')\delta(x-x')d^nx' \\ &= (2\pi\hbar)^n V(x,t')\end{aligned}$$

hence formula (5.34). □

There remains to show that the quantization rule (5.31)–(5.32) reduces to the Born–Jordan prescription

$$x^r p^s \longrightarrow \frac{1}{s+1}\sum_{\ell=0}^{s}\widehat{p}^{s-\ell}\widehat{x}^r\widehat{p}^{\ell} = \frac{1}{r+1}\sum_{j=0}^{r}\widehat{x}^{r-j}\widehat{p}^s\widehat{x}^j \tag{5.35}$$

when H is the monomial $x^r p^s$.

Proposition 3 *Let $H_{rs}(x,p) = x^r p^s$ where r and s are non-negative integers. Then the operator $\widehat{H_{rs}}$ defined by (5.31)–(5.32) is the Born–Jordan quantization of H_{rs}:*

$$\widehat{H_{rs}} = \frac{1}{r+1}\sum_{k=0}^{r}\widehat{x}^{r-k}\widehat{p}^s\widehat{x}^k. \tag{5.36}$$

Proof We have

$$\begin{aligned}\overline{H}_{rs}(x,x',p') &= \left(\int_0^1 (\tau x + (1-\tau)x')^r d\tau\right)p'^s \\ &= \sum_{k=0}^{r}\binom{r}{k}\left(\int_0^1 \tau^k(1-\tau)^{r-k}d\tau\right)x^k x'^{r-k}p'^s.\end{aligned}$$

As already noticed in Chap. 3, Sect. 3.2.4, the integral in the formula above is the beta function $B(k+1, r-k+1)$ and hence

$$\int_0^1 \tau^k(1-\tau)^{r-k}d\tau = \frac{k!(r-k)!}{(r+1)!}.$$

It follows that

$$\overline{H}_{rs}(x,x',p') = \frac{1}{r+1}\sum_{k=0}^{r}x^k x'^{r-k}p'^s$$

and the operator $\widehat{H_{rs}}$ is thus given by

$$\widehat{H_{rs}}\psi(x) = \left(\tfrac{1}{2\pi\hbar}\right)^n \frac{1}{r+1}\sum_{k=0}^{r} x^k \int e^{\frac{i}{\hbar}p'(x-x')} x'^{r-k} p'^s \psi(x') d^n p' d^n x'.$$

A straightforward calculation involving the Fourier inversion formula leads to

$$\left(\tfrac{1}{2\pi\hbar}\right)^n \int e^{\frac{i}{\hbar}p'(x-x')} x'^{r-k} p'^s \psi(x') d^n p' d^n x' \\ = (i\hbar)^s \frac{\partial^s}{\partial x^s}(x^{r-k}\psi(x))$$

and hence

$$\widehat{H_{rs}}\psi(x) = \frac{1}{r+1}\sum_{k=0}^{r} x^k (i\hbar)^s \frac{\partial^s}{\partial x^s}(x^{r-k}\psi(x))$$

which is precisely (5.36). □

We will revisit this summation over the parameter τ in Chap. 10. We point out that that the formula

$$\widehat{H}\psi(x) = \left(\tfrac{1}{2\pi\hbar}\right)^n \int e^{\frac{i}{\hbar}p'(x-x')} \overline{H}(x, x', p', t')\psi(x') d^n p' d^n x' \tag{5.37}$$

where the averaged Hamiltonian $\overline{H}$ is given by

$$\overline{H}(x, x', p', t') = \int_0^1 H(\tau x + (1-\tau)x', p', t') d\tau \tag{5.38}$$

obtained in Proposition 1 is at the heart of the pseudo-differential theory of Born–Jordan quantization we will study in Chap. 10.

References

1. M. de Gosson, *The Principles of Newtonian and Quantum Mechanics: The Need for Planck's Constant, h (With a foreword by B Hiley)* (Imperial College Press, World Scientific, 2001)
2. M. de Gosson, *Symplectic Geometry and Quantum Mechanics*. Operator Theory: Advances and Applications. Advances in Partial Differential Equations, vol. 166 (Birkhäuser, Basel, 2006)
3. L. Hörmander, *The analysis of linear partial differential operators I. Grundl. Math. Wissenschaft.*, vol. 256 (Springer, 1983)
4. F.G. Mehler, Über die Entwicklung einer function von beliebig vielen Variabeln nach Laplaceschen Funktionen höherer Ordnung. J. für Reine und Angew. Math. (in German) **66**, 161–176 (1866)

5. A. Peres, Unperformed experiments have no results. Am. J. Phys. **46**(7), 745–747 (1978)
6. A. Shimony, *The Search for a Naturalistic World View*, vol. 1 (Cambridge University Press, 1993)
7. M.H. Stone, On one-parameter unitary groups in Hilbert space. Ann. Math. Second Ser. **33**(3), 643–648 (1932)
8. G.N. Watson, Notes on generating functions of polynomials: (2) Hermite polynomials. J. Lond. Math. Soc. **8**, 194–199 (1933)

Part II
Mathematical Aspects of Born–Jordan Quantization

Chapter 6
The Weyl Correspondence

The Weyl correspondence, or Weyl quantization, is well-known both in harmonic analysis and quantum mechanics. It is part of the wider Weyl–Wigner–Moyal theory, where an emphasis on phase space techniques is made. There are two compelling reasons to study Weyl quantization before Born–Jordan quantization: it is easy to express Born–Jordan operators in terms of Weyl operators, and Weyl quantization is a particular cases of the Shubin τ-theory of pseudo-differential operators which will be developed later in this book, and which leads to an alternative definition of Born–Jordan quantization via an averaging process. For complementary material on the Weyl correspondence we refer to de Gosson [4, 5], Wong [14] or, at a more mathematically advanced level, Hörmander [8]. Littlejohn [9] gives an excellent short introduction to the Weyl–Wigner–Moyal formalism for physicists.

6.1 Definitions and Basics

6.1.1 *Notation and Terminology*

We will use multi-index notation: $\alpha = (\alpha_1, ..., \alpha_n) \in \mathbb{N}^n$, $|\alpha| = \alpha_1 + \cdots + \alpha_n$, and for $x = (x_1, ..., x_n)$ we write $x^\alpha = x_1^{\alpha_1} \cdots x_n^{\alpha_n}$ and $\partial_x^\alpha = \partial_{x_1}^{\alpha_1} \cdots \partial_{x_n}^{\alpha_n}$ where $\partial_{x_j} = \partial/\partial x_j$. We denote by σ the standard symplectic form on the phase space $\mathbb{R}^{2n} \equiv \mathbb{R}^n \times \mathbb{R}^n$: by definition $\sigma(z, z') = Jz \cdot z'$ where

$$J = \begin{pmatrix} 0_{n\times n} & I_{n\times n} \\ -I_{n\times n} & 0_{n\times n} \end{pmatrix}$$

is the "standard symplectic matrix". Note that $J^2 = -I_{2n\times 2n}$. Writing the phase space variable as $z = (x, p)$ we have

$$\sigma(z, z') = px' - p'x.$$

M.A. de Gosson, *Born–Jordan Quantization*, Fundamental Theories
of Physics 182, DOI 10.1007/978-3-319-27902-2_6

We will use the notation $\widehat{x}_j$ for the operator of multiplication by x_j and $\widehat{p}_j = -i\hbar\partial/\partial x_j$. These operators satisfy Born's canonical commutation relations $[\widehat{x}_j, \widehat{p}_k] = i\hbar\delta_{jk}$.

The space of all C^∞ complex-valued functions ψ on $\mathbb{R}^n$ such that for all multi-indices α and β the function $x^\alpha\partial_x^\beta\psi$ is bounded is denoted by $\mathcal{S}(\mathbb{R}^n)$ (it is the Schwartz space of rapidly decreasing functions). It is a Fréchet space for the semi-norms $||\psi||_{\alpha,\beta} = \sup|x^\alpha\partial_x^\beta\psi|$. The dual $\mathcal{S}'(\mathbb{R}^n)$ of $\mathcal{S}(\mathbb{R}^n)$ is the space of tempered distributions.

The L^2 scalar product of two complex functions ψ, ϕ on $\mathbb{R}^n$ is

$$\langle\psi|\phi\rangle = \int \psi^*(x)\phi(x)d^n x;$$

the associated norm is $||\psi|| = \langle\psi|\psi\rangle^{1/2}$. Notice that we are using here the physicist's convention and notation: $\langle\psi|\phi\rangle$ is linear in ϕ and antilinear in ψ (this choice is consistent with Dirac's bra-ket formalism).

6.1.2 Traditional Definition of Weyl Operators (in Physics)

We begin by giving the formal definition of Weyl operators used by physicists; it goes back to Weyl's foundational work [12, 13]. Let a be a function on phase space $\mathbb{R}^{2n}$; assuming that the Fourier transform

$$Fa(x,p) = \left(\tfrac{1}{2\pi\hbar}\right)^n \int e^{-\frac{i}{\hbar}(x_0 x + p_0 p)} a(x_0, p_0) d^n p_0 d^n x_0 \tag{6.1}$$

and its inverse exist, we can write

$$a(x,p) = \left(\tfrac{1}{2\pi\hbar}\right)^n \int e^{\frac{i}{\hbar}(x_0 x + p_0 p)} Fa(x_0, p_0) d^n p_0 d^n x_0. \tag{6.2}$$

One then defines the Weyl operator $\widehat{A} = a(\widehat{x}, \widehat{p})$ by making the formal substitution $x \longmapsto \widehat{x}$, $p \longmapsto \widehat{p}$ in the formula above. Thus:

$$\widehat{A} = \left(\tfrac{1}{2\pi\hbar}\right)^n \int e^{\frac{i}{\hbar}(x_0\widehat{x} + p_0\widehat{p})} Fa(x_0, p_0) d^n p_0 d^n x_0; \tag{6.3}$$

here $x_0\widehat{x}$ and $p_0\widehat{p}$ should be understood as $x_{0,1}\widehat{x}_1 + \cdots + x_{0,n}\widehat{x}_n$ and $p_{0,1}\widehat{p}_1 + \cdots + p_{0,n}\widehat{p}_n$. Of course, by modern standards, this is not a priori a very satisfactory definition, unless one is able to give a meaning to the exponential

$$\widehat{M}(z_0) = e^{\frac{i}{\hbar}(x_0\widehat{x} + p_0\widehat{p})}. \tag{6.4}$$

The latter is sometimes called *Weyl's characteristic operator* [2], and is usually interpreted as the power series

$$\widehat{M}(z_0) = \sum_{k=0}^{\infty} \frac{1}{k!}\left(\frac{i}{\hbar}\right)^k (x_0\widehat{x} + p_0\widehat{p})^k \tag{6.5}$$

where each term of this sum is supposed to act as a partial differential operator (see Cohen's book [2] for a detailed study of the operator $\widehat{M}(z_0)$). Needless to say, the computational difficulties can become quite considerable because of the non-commutativity of the operators $\widehat{x}$ and $\widehat{p}$. Nevertheless, using the Baker–Campbell–Hausdorff formula

$$e^{A+B} = e^{-\frac{1}{2}[A,B]}e^A e^B = e^{\frac{1}{2}[A,B]}e^B e^A$$

valid for A and B such that $[A,[A,B]] = [B,[A,B]] = 0$, we have, choosing $A = x_0\widehat{x}/\hbar$ and $B = p_0\widehat{p}/\hbar$ and taking into account the canonical commutation relation $[\widehat{x},\widehat{p}] = i\hbar$,

$$\widehat{M}(z_0) = e^{-\frac{i}{2\hbar}p_0x_0}e^{\frac{i}{\hbar}p_0\widehat{p}}e^{\frac{i}{\hbar}x_0\widehat{x}} \tag{6.6}$$

or, equivalently,

$$\widehat{M}(z_0) = e^{\frac{i}{2\hbar}p_0x_0}e^{\frac{i}{\hbar}x_0\widehat{x}}e^{\frac{i}{\hbar}p_0\widehat{p}}. \tag{6.7}$$

Formula (6.7) allows us to "guess" what the action of $\widehat{M}(z_0)$ is like. Limiting ourselves to the case $n = 1$ for notational simplicity, let ψ be a real analytic function. Expanding $e^{ip_0\widehat{p}/\hbar}$ in a power series, we have, since $i\widehat{p}/\hbar = \partial_x$,

$$e^{\frac{i}{\hbar}p_0\widehat{p}}\psi(x) = \sum_{k=0}^{\infty}\frac{1}{k!}x_0^k\partial_x^k\psi(x)$$

which we recognize as the Taylor series of $\psi(x+x_0)$ at the point x. Since the operator $e^{ix_0\widehat{x}/\hbar}$ is just multiplication by $e^{ix_0x/\hbar}$ we thus have

$$\widehat{M}(z_0)\psi(x) = e^{\frac{i}{\hbar}(x_0x+\frac{1}{2}p_0x_0)}\psi(x+x_0). \tag{6.8}$$

We will see below that this formula can be very simply—and rigorously—recovered if one uses the more physically motivated notion of Heisenberg operator.

6.1.3 *Traditional Definition of Weyl Operators (in Mathematics)*

In the mathematical literature—especially in the early times of the theory of pseudo-differential operators [6, 8, 11]—it is customary to define the Weyl operator $\widehat{A}$ with symbol a by the integral expression

$$\widehat{A}\psi(x) = \left(\tfrac{1}{2\pi\hbar}\right)^n \int e^{\frac{i}{\hbar}p(x-y)} a(\tfrac{1}{2}(x+y), p)\psi(y) d^n p d^n y. \tag{6.9}$$

While this definition immediately connects the theory of Weyl operators to the more general theory of pseudo-differential operators, which are operators of the type

$$\widehat{A}\psi(x) = \left(\tfrac{1}{2\pi\hbar}\right)^n \int e^{\frac{i}{\hbar}p(x-y)} a(x, y, p)\psi(y) d^n p d^n y,$$

it has certain technical disadvantages because of convergence problems for the integral. The latter is defined, for instance, if one chooses $\psi \in \mathcal{S}(\mathbb{R}^n)$ (which is not a very stringent requirement) and if the symbol a decreases fast enough in the p variables, typically satisfying an estimate of the type

$$|a(x,p)| \leq C(x)(1+|p|)^{-n-\varepsilon}$$

for some $\varepsilon > 0$. When this is not the case, the integral in (6.9) can be viewed as a distributional bracket, but one has still to be careful when manipulating it. When the symbol belongs to some "nice" classes (e.g. the "Hörmander classes" $S^m_{\rho,\delta}$ [8]), there is another clever way of giving a meaning to the non-convergent integral in (6.9); it consists in viewing it as a so-called "oscillatory integral". We refer to the literature on pseudo-differential operators for this approach; see for instance Chazarain and Piriou [1], Hörmander [8].

It is not immediately obvious why the "physical" definition (6.3) should coincide with the definition (6.9). That this is indeed the case will follow from the constructions below.

6.2 Heisenberg and Grossmann–Royer Operators

Let us introduce two related unitary operators on $L^2(\mathbb{R}^n)$: the Heisenberg operator (sometimes called the Heisenberg–Weyl operator), and the Grossmann–Royer operator (sometimes called the parity operators).

6.2.1 Definition and Discussion

The Heisenberg operators are well-known objects from quantum mechanics: they are unitary operators on $L^2(\mathbb{R}^n)$ which can be used to define the Heisenberg group (see de Gosson [4], Littlejohn [9]); we will give a dynamical description of these operators below. The Heisenberg operators are closely related to the less known Grossmann–Royer operators, which were defined independently by Grossmann [7] and Royer [10] in the mid 1970s. Heisenberg and Grossmann–Royer operators are in a sense Fourier transforms of each other, as we will see below.

Definition 1 Let $z_0 = (x_0, p_0)$ and $\psi \in \mathcal{S}'(\mathbb{R}^n)$. The Heisenberg operator $\widehat{T}(z_0)$ is defined by

$$\widehat{T}(z_0)\psi(x) = e^{\frac{i}{\hbar}(p_0 x - \frac{1}{2}p_0 x_0)}\psi(x - x_0) \tag{6.10}$$

and the Grossmann–Royer operator by

$$\widehat{T}_{GR}(z_0)\psi(x) = e^{\frac{2i}{\hbar}p_0(x-x_0)}\psi(2x_0 - x). \tag{6.11}$$

The operator $\widehat{T}(z_0)$ translates wavefunctions while boosting their momentum, and the operator $\widehat{T}_{\mathrm{GR}}(z_0)$ is a reflection with respect to z_0; it is easy to check by a simple calculation that both operators are related by the formula

$$\widehat{T}_{\mathrm{GR}}(z_0) = \widehat{T}(z_0)R^{\vee}\widehat{T}(z_0)^{-1} \tag{6.12}$$

where $R^{\vee} = \widehat{T}_{\mathrm{GR}}(0)$ is the reflection $R^{\vee}\psi(x) = \psi(-x)$. It follows, in particular, that $\widehat{T}_{\mathrm{GR}}(z_0)$ is an involution:

$$\widehat{T}_{\mathrm{GR}}(z_0)\widehat{T}_{\mathrm{GR}}(z_0) = I. \tag{6.13}$$

One proves, using the Leibniz formula, that $\widehat{T}(z_0)$ and $\widehat{T}_{\mathrm{GR}}(z_0)$ both are continuous automorphisms of the Schwartz space $\mathcal{S}(\mathbb{R}^n)$ and that their inverses are

$$\widehat{T}(z_0)^{-1} = \widehat{T}(-z_0),\ \widehat{T}_{\mathrm{GR}}(z_0)^{-1} = \widehat{T}_{\mathrm{GR}}(z_0)$$

(the second formula following from (6.13)). By duality, these operators extend to continuous automorphisms of the space $\mathcal{S}'(\mathbb{R}^n)$ of tempered distributions. It is moreover clear that $\widehat{T}(z_0)$ and $\widehat{T}_{\mathrm{GR}}(z_0)$ are unitary on $L^2(\mathbb{R}^n)$:

$$||\widehat{T}(z_0)\psi|| = ||\widehat{T}_{\mathrm{GR}}(z_0)\psi|| = ||\psi||.$$

The Heisenberg operators do not commute; they satisfy the relations

$$\widehat{T}(z_0)\widehat{T}(z_1) = e^{\frac{i}{\hbar}\sigma(z_0,z_1)}\widehat{T}(z_1)\widehat{T}(z_0) \tag{6.14}$$

$$\widehat{T}(z_0 + z_1) = e^{-\frac{i}{2\hbar}\sigma(z_0,z_1)}\widehat{T}(z_0)\widehat{T}(z_1); \tag{6.15}$$

the first formula is often viewed as an "exponentiated" version of the canonical commutation relations. The Grossmann–Royer operators satisfy the product formula

$$\widehat{T}_{\mathrm{GR}}(z_0)\widehat{T}_{\mathrm{GR}}(z_1) = e^{-\frac{2i}{\hbar}\sigma(z_0,z_1)}\widehat{T}(2(z_0 - z_1)) \tag{6.16}$$

(see de Gosson [4, 5] for detailed proofs of these relations).

In the two following examples we show that the Grossmann–Royer and Heisenberg operators can be used to define in a simple but non-standard way two well-known objects from phase space quantum mechanics:

Example 2 **The Wigner transform**. Let $\psi \in \mathcal{S}(\mathbb{R}^n)$. By definition, the Wigner transform of ψ is the function $\operatorname{Wig}\psi \in \mathcal{S}(\mathbb{R}^{2n})$ defined by

$$\operatorname{Wig}\psi(z) = \left(\tfrac{1}{\pi\hbar}\right)^n \langle \widehat{T}_{GR}(z)\psi|\psi\rangle. \tag{6.17}$$

Using formula (6.11) the integral expression of the Wigner transform is

$$\operatorname{Wig}\psi(z) = \left(\tfrac{1}{2\pi\hbar}\right)^n \int e^{-\frac{i}{\hbar}py}\psi(x+\tfrac{1}{2}y)\psi^*(x-\tfrac{1}{2}y)d^n y. \tag{6.18}$$

Similarly:

Example 3 **The ambiguity function**. Let $\psi \in \mathcal{S}(\mathbb{R}^n)$; the function $\operatorname{Amb}\psi \in \mathcal{S}(\mathbb{R}^{2n})$ defined by

$$\operatorname{Amb}\psi(z) = \left(\tfrac{1}{2\pi\hbar}\right)^n \langle \widehat{T}(z)\psi|\psi\rangle \tag{6.19}$$

is called the (radar) ambiguity function. Its integral expression is

$$\operatorname{Amb}\psi(z) = \left(\tfrac{1}{2\pi\hbar}\right)^n \int e^{-\frac{i}{\hbar}py}\psi(y+\tfrac{1}{2}x)\psi^*(y-\tfrac{1}{2}x)d^n y. \tag{6.20}$$

The Wigner transform and the ambiguity function, and their extensions to pairs of functions, will be studied in detail in Chap. 7.

6.2.2 Symplectic Fourier Transform

There is another important relation between the operators $\widehat{T}(z_0)$ and $\widehat{T}_{\mathrm{GR}}(z_0)$. Let us first define the symplectic Fourier transform of a function or distribution on $\mathbb{R}^{2n}$.

Definition 4 The symplectic Fourier transform $a_\sigma = F_\sigma a$, of a function $a \in \mathcal{S}(\mathbb{R}^{2n})$ is defined by $F_\sigma a(z) = Fa(Jz)$ where $J = \begin{pmatrix} 0 & I \\ -I & 0 \end{pmatrix}$ is the standard symplectic matrix and F the ordinary Fourier transform on $\mathbb{R}^{2n}$. Explicitly:

$$F_\sigma a(z) = \left(\tfrac{1}{2\pi\hbar}\right)^n \int e^{-\frac{i}{\hbar}\sigma(z,z')}a(z')d^{2n}z' \tag{6.21}$$

(recall that $\sigma(z,z') = Jz\cdot z'$ is the standard symplectic form).

As the ordinary Fourier transform, F_σ can be extended to a continuous automorphism $\mathcal{S}'(\mathbb{R}^{2n}) \longrightarrow \mathcal{S}'(\mathbb{R}^{2n})$ whose restriction to $L^2(\mathbb{R}^{2n})$ is unitary. The usual Plancherel formula takes here the form

$$\int a(z)b^*(z)d^{2n}z = \int a_\sigma(z)(b_\sigma)^*(z)d^{2n}z \tag{6.22}$$

or, equivalently,

$$\int a(z)b(z)d^{2n}z = \int a_\sigma(z)b_\sigma(-z)d^{2n}z. \tag{6.23}$$

Moreover, since $F^2a(z) = a(-z)$ and $J^2 = -I$ the symplectic Fourier transform is involutive: $F_\sigma^2 = F_\sigma$ and is hence equal to its own inverse: $F_\sigma^{-1} = F_\sigma$. We note that the symplectic Fourier transform satisfies the usual convolution formulas

$$F_\sigma(a * b) = (2\pi\hbar)^n(F_\sigma a)(F_\sigma b) \tag{6.24}$$

and

$$F_\sigma(ab) = (2\pi\hbar)^{-n}F_\sigma a * F_\sigma b. \tag{6.25}$$

Proposition 5 *Let $\psi \in \mathcal{S}(\mathbb{R}^n)$ (or, more generally, $\psi \in \mathcal{S}'(\mathbb{R}^n)$). We have*

$$\widehat{T}_{GR}(z_0)\psi(x) = 2^{-n}F_\sigma[\widehat{T}(\cdot)\psi(x)](-z_0) \tag{6.26}$$

and

$$\widehat{T}(z_0)\psi(x) = 2^nF_\sigma[\widehat{T}_{GR}(\cdot)\psi(x)](-z_0). \tag{6.27}$$

Proof (Cf. de Gosson [8], Ch. 8). Formula (6.27) is equivalent to formula (6.26) since the symplectic Fourier transform is involutive. To prove (6.26) it suffices to consider the case $\psi \in \mathcal{S}(\mathbb{R}^n)$ (the general case follows by duality and density). We thus have to show that

$$\widehat{T}_{\mathrm{GR}}(z_0)\psi(x) = \left(\tfrac{1}{4\pi\hbar}\right)^n \int e^{\frac{i}{\hbar}\sigma(z_0,z')}\widehat{T}(z')\psi(x)d^{2n}z'. \tag{6.28}$$

Denoting by $A(x)$ the right-hand side of (6.28) we have, using the definition of the symplectic Fourier transform,

$$\begin{aligned}A(x) &= \left(\tfrac{1}{4\pi\hbar}\right)^n \int e^{\frac{i}{\hbar}\sigma(z_0,z')}\widehat{T}(z')\psi(x)d^{2n}z' \\ &= \left(\tfrac{1}{4\pi\hbar}\right)^n \int e^{\frac{i}{\hbar}(p_0x'-p'x_0+p'x-\frac{1}{2}p'x')}\psi(x-x')d^{2n}z' \\ &= \left(\tfrac{1}{4\pi\hbar}\right)^n \int \left(\int e^{\frac{i}{\hbar}p'(x-x_0-\frac{1}{2}x')}d^np'\right)e^{\frac{i}{\hbar}p_0x'}\psi(x)d^nx'.\end{aligned}$$

Now, $F(1) = (2\pi\hbar)^n\,\delta$, and hence

$$\int e^{\frac{i}{\hbar}p'(x-x_0-\frac{1}{2}x')}d^n p' = (2\pi\hbar)^n\,\delta(x - x_0 - \tfrac{1}{2}x').$$

Setting $y = \frac{1}{2}x'$ we get,

$$\begin{aligned} A(x) &= 2^{-n}\int \delta(x - x_0 - \tfrac{1}{2}x')e^{\frac{i}{\hbar}p_0 x'}\psi(x)d^n x' \\ &= \int \delta(y + x_0 - x)e^{\frac{2i}{\hbar}p_0 y}\psi(x)d^n y \\ &= e^{\frac{2i}{\hbar}p_0(x-x_0)}\psi(-x + 2x_0) \\ &= \widehat{T}_{\mathrm{GR}}(z_0)\psi(x) \end{aligned}$$

which proves formula (6.26). ■

An immediate consequence is that the Wigner distribution and the ambiguity function introduced in the Examples 2 and 3 above are symplectic Fourier transforms of each other:

Corollary 6 *Let* $\psi \in L^2(\mathbb{R}^n)$. *We have* $\mathrm{Amb}\,\psi = F_\sigma \mathrm{Wig}\,\psi$ *and* $\mathrm{Wig}\,\psi = F_\sigma \mathrm{Amb}\,\psi$.

Proof It is a straightforward consequence of the Proposition above, using the definitions (6.17) and (6.19) of the Wigner and ambiguity functions. ■

6.2.3 Dynamical Interpretation of the Heisenberg Operator

It is not uninteresting to give a dynamical interpretation of the Heisenberg operator. Consider, for fixed $z_0 \in \mathbb{R}^{2n}$, the function

$$H_0(z) = \sigma(z, z_0). \tag{6.29}$$

We call this function the "displacement Hamiltonian": the flow (f_t^0) determined by the Hamilton equations $\dot{z} = J\partial_z H_0(z)$ consists of the phase space translations $f_t^0 : z \longmapsto z + tz_0$. Consider now the operator

$$\widehat{H}_0 = \sigma(\widehat{z}, z_0) = x_0\widehat{p} - p_0\widehat{x} \tag{6.30}$$

obtained by formally replacing $z = (x, p)$ with $\widehat{z} = (\widehat{x}, \widehat{p})$ where $\widehat{x}_j$ is multiplication by x_j and $\widehat{p}_j = -i\hbar\partial_{x_j}$: explicitly

$$\widehat{H}_0 = \sum_{j=1}^{n} x_{0,j}\widehat{p}_j - p_{0,j}\widehat{x}_j$$

which we write $\widehat{H}_0 = x_0\widehat{p} - p_0\widehat{x}$. The solution of the corresponding Schrödinger's equation

$$i\hbar\partial_t\psi = \widehat{H}_0\psi,\ \psi(\cdot, 0) = \psi_0 \tag{6.31}$$

is formally given by the formula

$$\psi(x,t) = e^{-\frac{i}{\hbar}\widehat{H}_0 t}\psi_0(x) = e^{-\frac{i}{\hbar}\sigma(\widehat{z},z_0)t}\psi_0(x). \tag{6.32}$$

One verifies by a direct calculation that one has explicitly

$$\psi(x,t) = e^{\frac{i}{\hbar}(tp_0x - \frac{1}{2}t^2 p_0x_0)}\psi_0(x - tx_0). \tag{6.33}$$

Thus, $\widehat{T}(z_0)\psi_0$ is the time-one solution of the Schrödinger equation (6.31). In operator notation:

$$\widehat{T}(z_0) = e^{-\frac{i}{\hbar}\sigma(\widehat{z},z_0)}. \tag{6.34}$$

Recall that we defined above the Weyl characteristic operator as being the operator

$$\widehat{M}(z_0) = e^{\frac{i}{\hbar}(p_0\widehat{x} + x_0\widehat{p})}.$$

This operator is closely related to the Heisenberg operator function $\widehat{T}(z_0)$—it is in fact just a variant thereof: since $\sigma(J\widehat{z}, z_0) = x_0\widehat{p} + p_0\widehat{x}$ we have the simple relation

$$\widehat{M}(z_0) = e^{\frac{i}{\hbar}(x_0\widehat{x} + p_0\widehat{p})} = \widehat{T}(-Jz_0). \tag{6.35}$$

The properties of $\widehat{M}(z_0)$, including commutation relations, immediately follow from those of $\widehat{T}(z_0)$. Using the definition (6.10) of $\widehat{T}(z_0)$ we recover the action (6.8) of $\widehat{M}(z_0)$ on functions or distributions. In fact, changing (x_0, p_0) into $(-p_0, x_0)$, we get

$$\widehat{M}(z_0)\psi(x) = e^{\frac{i}{\hbar}(x_0x + \frac{1}{2}p_0x_0)}\psi(x + p_0).$$

Notice that we did not have to make any analyticity assumption on ψ.

6.3 Weyl Operators: Harmonic Analysis

We are now able to define Weyl operators in a rigorous and convenient way.

6.3.1 Definition of a Weyl Operator

Let $a \in \mathcal{S}'(\mathbb{R}^{2n})$, hereafter to be called "symbol" (or "classical observable[1]", in the physical literature). We denote by $\langle\langle \cdot, \cdot \rangle\rangle$ the distributional bracket on $\mathbb{R}^{2n}$: if U is a distribution on $\mathbb{R}^{2n}$ and ϕ a test function, then the complex number $\langle\langle U, \phi \rangle\rangle$ is the value of the functional U at ϕ. In our case we will mostly have $U \in \mathcal{S}'(\mathbb{R}^{2n})$ and $\phi \in \mathcal{S}(\mathbb{R}^{2n})$; as usual we will often abuse notation by writing

$$\langle\langle U, \phi \rangle\rangle = \int U(z)\phi(z)d^{2n}z.$$

Definition 7 The linear operator $\widehat{A} : \mathcal{S}(\mathbb{R}^n) \longrightarrow \mathcal{S}'(\mathbb{R}^n)$ defined by

$$\widehat{A}\psi = \left(\tfrac{1}{\pi\hbar}\right)^n \langle\langle a(\cdot), \widehat{T}_{GR}(\cdot)\psi \rangle\rangle \tag{6.36}$$

is the Weyl operator with symbol a (or the pseudo-differential operator with Weyl symbol a). In integral form:

$$\widehat{A}\psi(x) = \left(\tfrac{1}{\pi\hbar}\right)^n \int a(z_0)\widehat{T}_{\mathrm{GR}}(z_0)\psi(x)d^{2n}z_0. \tag{6.37}$$

We will write $\widehat{A} = \mathrm{Op_W}(a)$ and sometimes use the notation $\widehat{A} \overset{Weyl}{\longleftrightarrow} a$ or $a \overset{Weyl}{\longleftrightarrow} \widehat{A}$.

The right-hand side of (6.36) makes sense since for every $z_0 \in \mathbb{R}^{2n}$ the function $\widehat{T}_{\mathrm{GR}}(z_0)\psi$ is in $\mathcal{S}(\mathbb{R}^{2n})$; this ensures that $\langle\langle a(\cdot), \widehat{T}_{\mathrm{GR}}(\cdot)\psi \rangle\rangle$ is defined. We have chosen as natural domain for a Weyl operator the Schwartz space $\mathcal{S}(\mathbb{R}^n)$; in many cases $\widehat{A}$ can be extended to larger classes of functions.

The following result connects the definition above with the traditional integral definition (6.9) given above:

Proposition 8 *If $a \in \mathcal{S}(\mathbb{R}^{2n})$ and $\psi \in \mathcal{S}(\mathbb{R}^n)$ then*

$$\widehat{A}\psi(x) = \left(\tfrac{1}{2\pi\hbar}\right)^n \int e^{\frac{i}{\hbar}p(x-y)}a(\tfrac{1}{2}(x+y), p)\psi(y)d^n p d^n y. \tag{6.38}$$

Proof Formula (6.37) yields, using the definition (6.11) of the Grossmann–Royer operator,

$$\widehat{A}\psi(x) = \left(\tfrac{1}{\pi\hbar}\right)^n \int a(z_0)e^{\frac{2i}{\hbar}p_0(x-x_0)}\psi(2x-x_0)d^{2n}z_0.$$

[1] Strictly speaking, we should reserve the term "observable" to *real* symbols; we are thus committing a slight abuse of terminology.

Setting $y = 2x - x_0$ and $p = p_0$ we get

$$\widehat{A}\psi(x) = \left(\tfrac{1}{2\pi\hbar}\right)^n \int e^{\frac{i}{\hbar}p(x-y)} a(\tfrac{1}{2}(x+y), p)\psi(y)dydp$$

which is precisely formula (6.38). ∎

6.3.2 Harmonic Analysis

Weyl operators can alternatively be defined in terms of the Heisenberg operator (and this redefinition has many advantages, often leading to simpler formulas):

Proposition 9 *Let $a \in \mathcal{S}'(\mathbb{R}^{2n})$, $\widehat{A} = \mathrm{Op_W}(a)$ and $\psi \in \mathcal{S}(\mathbb{R}^n)$. (i) We have*

$$\widehat{A}\psi = \left(\tfrac{1}{2\pi\hbar}\right)^n \langle\langle a_\sigma(\cdot), \widehat{T}(\cdot)\psi\rangle\rangle \tag{6.39}$$

where $a_\sigma = F_\sigma a$ is the symplectic Fourier transform of a; in integral notation

$$\widehat{A}\psi(x) = \left(\tfrac{1}{2\pi\hbar}\right)^n \int a_\sigma(z_0)\widehat{T}(z_0)\psi(x)d^{2n}z_0. \tag{6.40}$$

Proof It is sufficient to prove (6.39) for $a \in \mathcal{S}(\mathbb{R}^n)$; the general case follows by a density argument. We have, using successively Parseval's formula and the relation (6.27),

$$\begin{aligned}\langle\langle a(\cdot), \widehat{T}_{\mathrm{GR}}(\cdot)\psi(x)\rangle\rangle &= \int a_\sigma(z_0)F_\sigma[\widehat{T}_{\mathrm{GR}}(\cdot)\psi(x)](-z_0)d^{2n}z_0 \\ &= 2^{-n}\int a_\sigma(z_0)\widehat{T}(z_0)\psi(x)d^{2n}z_0 \\ &= 2^{-n}\langle\langle a_\sigma(\cdot), \widehat{T}(\cdot)\psi(x)\rangle\rangle\end{aligned}$$

hence the formulas (6.40) and (6.39). ∎

Since we will use quite often a_σ in this, and the forthcoming chapters let us give it a name:

Definition 10 The function $a_\sigma = F_\sigma a$ is called the *covariant* (or "twisted") symbol of $\widehat{A} = \mathrm{Op_W}(a)$.

Recall formula (6.34) that the Heisenberg operator can be written $\widehat{T}(z_0) = e^{-i\sigma(\widehat{z},z_0)/\hbar}$. This suggests that $\widehat{T}(z_0)$ is just the Weyl quantization of the function $e^{-i\sigma(z,z_0)/\hbar}$. This is indeed the case:

Proposition 11 *The operator with Weyl symbol $a_0(z) = e^{-i\sigma(z,z_0)/\hbar}$ is the Heisenberg–Weyl operator $\widehat{T}(z_0)$. Equivalently, the covariant symbol of $\widehat{T}(z_0)$ is $(2\pi\hbar)^n\delta(z - z_0)$.*

Proof Let us set $a_0(z) = e^{-i\sigma(z,z_0)/\hbar}$ and $\widehat{A}_0 = \operatorname{Op_W}(a_0)$. We have

$$\widehat{A}_0 = \left(\tfrac{1}{2\pi\hbar}\right)^n \int (a_0)_\sigma(z)\widehat{T}(z)d^{2n}z$$

where the covariant symbol $(a_0)_\sigma$ is given by

$$\begin{aligned}(a_0)_\sigma(z) &= \left(\tfrac{1}{2\pi\hbar}\right)^n \int e^{-\frac{i}{\hbar}\sigma(z,z')}e^{-\frac{i}{\hbar}\sigma(z',z_0)}d^{2n}z' \\ &= \left(\tfrac{1}{2\pi\hbar}\right)^n \int e^{-\frac{i}{\hbar}\sigma(z-z_0,z')}d^{2n}z' \\ &= (2\pi\hbar)^n\delta(z-z_0)\end{aligned}$$

(the last equality because $F_\sigma(1) = (2\pi\hbar)^n\delta$); hence

$$\widehat{A}_0 = \int \delta(z-z_0)\widehat{T}(z)d^{2n}z = \widehat{T}(z_0)$$

taking into account the fact that $\delta(z-z_0) = \delta(z_0-z)$. ■

6.3.3 The Kernel of a Weyl Operator

The bijectivity of a Weyl operator is most easily proven using its (distributional) kernel, since it allows us to invoke Schwartz's kernel theorem. It also allows the study of boundedness properties of Weyl operators on the Hilbert space $L^2(\mathbb{R}^n)$.

Every linear operator $A : \mathcal{S}(\mathbb{R}^n) \longrightarrow \mathcal{S}'(\mathbb{R}^n)$ has a kernel $K \in \mathcal{S}'(\mathbb{R}^n \times \mathbb{R}^n)$ provided it is continuous (in the weak $*$-star topology) from $\mathcal{S}(\mathbb{R}^n)$ to $\mathcal{S}'(\mathbb{R}^n)$; this is Schwartz's kernel theorem briefly discussed in Sect. 5.2.1 of Chap. 5. The Schwartz kernel is a distribution $K \in \mathcal{S}'(\mathbb{R}^n \times \mathbb{R}^n)$ such that

$$\langle A\psi, \phi\rangle = \langle\langle K, \phi\otimes\psi\rangle\rangle$$

for all $\phi, \psi \in \mathcal{S}(\mathbb{R}^n)$; as usual $\langle\langle\cdot,\cdot\rangle\rangle$ is the distributional bracket on the space $\mathbb{R}^n \times \mathbb{R}^n$. In integral notation:

$$A\psi(x) = \int K(x,y)\psi(y)d^n y. \tag{6.41}$$

The kernel K and the symbol a of a Weyl operator are related by two simple formulas:

Proposition 12 *Let* $\widehat{A} = \operatorname{Op_W}(a)$ *have symbol* $a \in \mathcal{S}'(\mathbb{R}^{2n})$ *and kernel* $K \in \mathcal{S}'(\mathbb{R}^n \times \mathbb{R}^n)$. *(i) We have*

$$K(x,y) = \left(\tfrac{1}{2\pi\hbar}\right)^{n/2} (F_2^{-1}a)(\tfrac{1}{2}(x+y), x-y) \tag{6.42}$$

where F_2^{-1} is the inverse Fourier transform in the second set of variables; in integral form:

$$K(x,y) = \left(\tfrac{1}{2\pi\hbar}\right)^{n} \int e^{\frac{i}{\hbar}p(x-y)} a(\tfrac{1}{2}(x+y), p) d^n p; \tag{6.43}$$

(ii) Conversely

$$a(x,p) = \int e^{-\frac{i}{\hbar}py} K(x+\tfrac{1}{2}y, x-\tfrac{1}{2}y) d^n y. \tag{6.44}$$

Proof (i) Formulas (6.42) and (6.43) are obvious, comparing (6.41) with the integral representation (6.38) of $\widehat{A}$. (ii) Formula (6.44) follows from (6.43) or (6.42) using the Fourier inversion formula. ■

We have been calling the association $a \overset{\text{Weyl}}{\longleftrightarrow} \widehat{A}$ between a symbol and the corresponding Weyl operator "Weyl correspondence". This terminology is justified by the following result which shows that $a \overset{\text{Weyl}}{\longleftrightarrow} \widehat{A}$ is indeed a bijection.

We denote by $\mathcal{L}(\mathcal{S}(\mathbb{R}^n), \mathcal{S}'(\mathbb{R}^n))$ the vector space of all continuous linear mappings $\mathcal{S}(\mathbb{R}^n) \longrightarrow \mathcal{S}'(\mathbb{R}^n)$.

Proposition 13 *The mapping*

$$\mathrm{Op_W} : \mathcal{S}'(\mathbb{R}^{2n}) \longrightarrow \mathcal{L}(\mathcal{S}(\mathbb{R}^n), \mathcal{S}'(\mathbb{R}^n))$$

is a vector space isomorphism. More precisely: (i) The Weyl correspondence $a \overset{\text{Weyl}}{\longleftrightarrow} \widehat{A}$ is linear and injective: If $a \overset{\text{Weyl}}{\longleftrightarrow} \widehat{A}$ and $a' \overset{\text{Weyl}}{\longleftrightarrow} \widehat{A}$ then $a = a'$. In particular $1 \overset{\text{Weyl}}{\longleftrightarrow} I_{\mathrm{d}}$ (the identity operator on $\mathcal{S}'(\mathbb{R}^n)$). (ii) Every $A \in \mathcal{L}(\mathcal{S}(\mathbb{R}^n), \mathcal{S}'(\mathbb{R}^n))$ can be written $A = \mathrm{Op_W}(a)$ for some (unique) $a \in \mathcal{S}'(\mathbb{R}^{2n})$.

Proof (i) The linearity is obvious. To establish the injectivity it suffices to show that if $\widehat{A} \overset{\text{Weyl}}{\longleftrightarrow} a$ and $\widehat{A} = I_{\mathrm{d}}$ then $a = 1$. In view of (6.41) the condition $\widehat{A} = I_{\mathrm{d}}$ is equivalent to $K(x,y) = \delta(x-y)$, which requires (formula (6.43) that $a(\frac{1}{2}(x+y), p) = 1$ for all x, p that is $a = 1$. That $1 \overset{\text{Weyl}}{\longleftrightarrow} I_{\mathrm{d}}$ is clear using a similar argument. (ii) The surjectivity of the correspondence $a \overset{\text{Weyl}}{\longleftrightarrow} \widehat{A}$ is obvious in view of the discussion above, using Schwartz's kernel theorem: every operator $A \in \mathcal{L}(\mathcal{S}(\mathbb{R}^n), \mathcal{S}'(\mathbb{R}^n))$ has a kernel K, and its Weyl symbol a is then given by formula (6.44); the uniqueness of a follows from the injectivity property proven in (i). ■

6.4 Two Examples

6.4.1 *The Weyl Quantization of Monomials*

Physically speaking, any realistic "quantization" of classical observables should satisfy the Schrödinger prescription: to a symbol $a(z) = x_j p_j$ should correspond the operator

$$\widehat{A} = \tfrac{1}{2}(\widehat{x_j}\widehat{p_j} + \widehat{p_j}\widehat{x_j}). \tag{6.45}$$

Let us check that the Weyl correspondence $\widehat{A} \overset{\text{Weyl}}{\longleftrightarrow} a$ satisfies this requirement. We have, taking $a(z) = x_j p_j$ and hence

$$\begin{aligned}(2\pi\hbar)^n \widehat{A}\psi(x) &= \tfrac{1}{2}\int e^{\frac{i}{\hbar}p(x-y)} x_j p_j \psi(y) d^n y d^n p \\ &+ \tfrac{1}{2}\int e^{\frac{i}{\hbar}p(x-y)} y_j p_j \psi(y) d^n y d^n p.\end{aligned}$$

Formula (6.45) follows in view of the obvious equalities

$$\begin{aligned}&\left(\tfrac{1}{2\pi\hbar}\right)^n x_j \int e^{\frac{i}{\hbar}p(x-y)} p_j \psi(y) d^n y d^n p = x_j \widehat{p}\psi(x) \\ &\left(\tfrac{1}{2\pi\hbar}\right)^n \int e^{\frac{i}{\hbar}p(x-y)} p_j y_j \psi(y) d^n y d^n p = \widehat{p}(x_j\psi)(x).\end{aligned}$$

More generally, a similar argument shows that the Weyl quantization of an arbitrary monomial $x^r p^s$ leads to the Weyl–McCoy formula (2.11):

$$\mathrm{Op}_W(x^r p^s) = \frac{1}{2^s}\sum_{\ell=0}^{s}\binom{s}{\ell}\widehat{p}^{s-\ell}\widehat{x}^r\widehat{p}^{\ell}. \tag{6.46}$$

We will not perform the detailed calculations here since the result is a particular case of the more general rule obtained using Shubin's τ-pseudo-differential calculus which we will study later on.

6.4.2 *Physical Hamiltonians*

Assume that H is a (possibly time-dependent) function of the type

$$H(z,t) = \sum_{j=1}^{n} \frac{1}{2m_j}(p_j - \mathcal{A}_j(x,t))^2 + V(x,t); \tag{6.47}$$

where the potentials $\mathcal{A} = (\mathcal{A}_1, ..., \mathcal{A}_n)$ and V are twice continuously differentiable in the x variables, and continuous in t. Let us show that $\widehat{H} = \mathrm{Op}(H)$ is the usual quantum operator

$$\widehat{H} = \sum_{j=1}^{n} \frac{1}{2m_j}\left(-i\hbar\frac{\partial}{\partial x_j} - \mathcal{A}_j(x,t)\right)^2 + V(x,t). \tag{6.48}$$

Since $\mathrm{Op}(p_j^2) = \widehat{p}_j^2$, $\mathrm{Op}(\mathcal{A}_j) = \mathcal{A}_j$ and $\mathrm{Op}(V) = V$, we only need to bother about the cross-products $p_j\mathcal{A}j$. Dropping the index j and assuming $n = 1$ we claim that

$$\mathrm{Op}(p\mathcal{A})\psi) = -\frac{i\hbar}{2}\left[\frac{\partial}{\partial x}(\mathcal{A}\psi) + \mathcal{A}\frac{\partial\psi}{\partial x}\right], \tag{6.49}$$

from which (6.48) readily follows. Let us prove (10.18); it is sufficient to assume $n = 1$. We have (formula (6.38))

$$\begin{aligned}\mathrm{Op}(p\mathcal{A})\psi(x) &= \frac{1}{2\pi\hbar}\int e^{\frac{i}{\hbar}p(x-y)}p\mathcal{A}(\tfrac{1}{2}(x+y),t)\psi(y)dydp \\ &= \int_{-\infty}^{\infty}\left[\frac{1}{2\pi\hbar}\int_{-\infty}^{\infty} e^{\frac{i}{\hbar}p(x-y)}pdp\right]\mathcal{A}(\tfrac{1}{2}(x+y),t)\psi(x')dx'.\end{aligned}$$

In view of the Fourier inversion formula the expression between the square brackets is $-i\hbar\delta'(x-y)$ so that

$$\begin{aligned}\mathrm{Op}(p\mathcal{A})\psi(x) &= -i\hbar\int_{-\infty}^{\infty}\delta'(x-y)\mathcal{A}(\tfrac{1}{2}(x+y),t)\psi(y)dy \\ &= -i\hbar\int_{-\infty}^{\infty}\delta(x-y)\frac{\partial}{\partial y}(\mathcal{A}(\tfrac{1}{2}(x+y),t)\psi(y))dy \\ &= \frac{-i\hbar}{2}\left(\frac{\partial\mathcal{A}}{\partial x}(x,t)\psi(x) + \mathcal{A}(x,t)\frac{\partial\psi}{\partial x}(x))\right)\end{aligned}$$

which is formula (6.49).

In many physics texts quantization formulas of the type (6.48) are formally derived using the monomial rule (6.46): one expands the vector potential $\mathcal{A}$ in a Taylor series at $x = 0$ and one then quantizes separately each term in the series obtained for the product $p_j\mathcal{A}$. Such a procedure is of course only mathematically legitimate if one assumes that the vector potential $\mathcal{A}$ is (a real) analytic function.

6.5 L^2-Boundedness of Weyl Operators

The following result is a rather elementary result giving a necessary and sufficient condition for the L^2-boundedness of a Weyl operator:

Proposition 14 *Let $\widehat{A} = \mathrm{Op_W}(a)$ have kernel K. (i) We have $a \in L^2(\mathbb{R}^{2n})$ if and only if $K \in L^2(\mathbb{R}^n \times \mathbb{R}^n)$; (ii) when this is the case $\widehat{A}$ is a bounded operator on $L^2(\mathbb{R}^n)$.*

Proof (Cf. [5], Sect. 10.1). (i) Let us prove that

$$||a|| = (2\pi\hbar)^{n/2} \,|||K|||$$

($|||\cdot|||$ the L^2 norm on $\mathbb{R}^{2n}$). In view of formula (6.44) the symbol a is, for fixed x, $(2\pi\hbar)^{n/2}$ times the Fourier transform of the partial function $y \longmapsto K(x+\frac{1}{2}y, x-\frac{1}{2}y)$ hence, by Plancherel's formula,

$$\int |a(x,p)|^2 d^n p = (2\pi\hbar)^n \int |K(x+\tfrac{1}{2}y, x-\tfrac{1}{2}y)|^2 d^n y. \tag{6.50}$$

Integrating this equality with respect to x we get

$$\begin{aligned}\int |a(x,p)|^2 d^n p d^n x &= (2\pi\hbar)^n \int \left(\int |K(x+\tfrac{1}{2}y, x-\tfrac{1}{2}y)|^2 d^n y \right) d^n x \\ &= (2\pi\hbar)^n \int |K(x+\tfrac{1}{2}y, x-\tfrac{1}{2}y)|^2 d^n x d^n y\end{aligned}$$

where we have applied Fubini's theorem. Making the change of variables $x' = x+\frac{1}{2}y$ and $y' = x - \frac{1}{2}y$ we have $d^n x' d^n y' = d^n x d^n y$ hence

$$\int |a(x,p)|^2 d^n p d^n x = (2\pi\hbar)^n \int |K(x', y')|^2 d^n x' d^n y'$$

and we thus have $a \in L^2(\mathbb{R}^{2n})$ if and only if $K \in L^2(\mathbb{R}^n \times \mathbb{R}^n)$. (ii) The L^2-boundedness of $\widehat{A}$ immediately follows, using the Cauchy-Schwarz inequality: writing

$$\widehat{A}\psi(x) = \int K(x,y)\psi(y) d^n y$$

we have

$$|\widehat{A}\psi(x)|^2 \leq \int |K(x,y)|^2 d^n y \int |\psi(y)|^2 d^n y$$

and hence

$$||\widehat{A}\psi|| = \int |\widehat{A}\psi(x)|^2 dx \leq |||K||| ||\psi||^2$$

where $|||K|||$ is the L^2 norm of the kernel K. ∎

Notice that the condition $a \in L^2(\mathbb{R}^{2n})$ in the result above can be replaced with $a_\sigma \in L^2(\mathbb{R}^{2n})$ since the symplectic Fourier transform is unitary: $|||a||| = |||a_\sigma|||$, and in this case we have

$$|||a_\sigma||| = (2\pi\hbar)^{n/2}\,|||K|||. \tag{6.51}$$

The condition stated in the proposition above is sufficient, but not necessary. In fact, if we assume that the Fourier transform of the symbol is absolutely integrable we again have L^2-boundedness:

Proposition 15 *Let $\widehat{A} = \mathrm{Op_W}(a)$. If $a_\sigma \in L^1(\mathbb{R}^{2n})$ then $\widehat{A}$ is bounded on $L^2(\mathbb{R}^n)$ and the operator norm of $\widehat{A}$ on $L^2(\mathbb{R}^n)$ satisfies*

$$||\widehat{A}|| \leq \left(\tfrac{1}{2\pi\hbar}\right)^{2n} ||a_\sigma||_{L^1(\mathbb{R}^{2n})}. \tag{6.52}$$

Proof Recall that the covariant symbol a_σ is defined by $a_\sigma(z) = F_\sigma a(z)$. In view of the density of $\mathcal{S}(\mathbb{R}^n)$ in $L^2(\mathbb{R}^n)$ it suffices to assume that $\psi \in \mathcal{S}(\mathbb{R}^n)$. Let us prove the inequality (6.52); the boundedness of $\widehat{A}$ will follow. Rewriting formula (6.42) as

$$K(x, y) = \left(\tfrac{1}{2\pi\hbar}\right)^{n/2} (F_2 a)(\tfrac{1}{2}(x + y), y - x),$$

let $Fa = (F_1 \otimes F_2)a$ be the Fourier transform of a; using the Fourier inversion formula we have

$$F_2 a(\tfrac{1}{2}(x + y), y - x) = \left(\tfrac{1}{2\pi\hbar}\right)^{n/2} \int e^{\frac{i}{2\hbar}(x+y)u} F(u, y - x) d^n u$$

and hence

$$K(x, y) = \left(\tfrac{1}{2\pi\hbar}\right)^{n} \int e^{\frac{i}{2\hbar}(x+y)u} Fa(u, y - x) d^n u.$$

It follows that

$$\int |K(x, y)| dx \leq \left(\tfrac{1}{2\pi\hbar}\right)^{n} \int |Fa(u, y - x)| d^n u d^n x$$

$$\int |K(x, y)| dy \leq \left(\tfrac{1}{2\pi\hbar}\right)^{n} \int |Fa(u, y - x)| d^n u d^n y;$$

equivalently, setting $\eta = y - x$,

$$\int |K(x,y)|d^n x \leq \left(\tfrac{1}{2\pi\hbar}\right)^n ||Fa||_{L^1(\mathbb{R}^{2n})}$$
$$\int |K(x,y)|d^n y \leq \left(\tfrac{1}{2\pi\hbar}\right)^n ||Fa||_{L^1(\mathbb{R}^{2n})}.$$

Set now $C = (2\pi\hbar)^{-n}||a_\sigma||_{L^1}$; by Cauchy–Schwarz's inequality we have

$$\begin{aligned}|\widehat{A}\psi(x)|^2 &\leq \int |K(x,y)|d^n y \int |K(x,y)||\psi(y)|^2 d^n y \\ &\leq C^2 \int |K(x,y)||\psi(y)|^2 d^n y\end{aligned}$$

and hence

$$\int |\widehat{A}\psi(x)|^2 d^n x \leq C^2 \int \left(\int |K(x,y)|dx \right) |\psi(y)|^2 d^n y$$

that is

$$\int |\widehat{A}\psi(x)|^2 d^n x \leq C^2 \int |\psi(y)|^2 d^n y$$

which is the estimate (6.52). ■

The study of regularity properties of Weyl operators on $L^2(\mathbb{R}^n)$ has become something of an industry since the early 1990's. Cordes [3] has proven the following rather general property:

Proposition 16 *Assume that the symbol a satisfies the conditions* $\partial_x^\alpha \partial_p^\beta a \in L^\infty(\mathbb{R}^{2n})$ *for all multi-indices* α, β *such that* $|\alpha|, |\beta| \leq [n/2]+1$. *Then the operator* $\widehat{A} \overset{\text{Weyl}}{\longleftrightarrow} a$ *is bounded on the space* $L^2(\mathbb{R}^n)$.

The proof of this result is rather technical, so we will not reproduce it here. It implies in particular that every $a \in C^\infty(\mathbb{R}^{2n})$ which is, together with all its derivatives, bounded determines a bounded operator on $L^2(\mathbb{R}^n)$.

6.6 Adjoints and Products

We determine the symbol of the formal adjoint of a Weyl operator. We also calculate the symbol of the product of two Weyl operators.

6.6.1 The (Formal) Adjoint of a Weyl Operator

The formal adjoint $\widehat{A}^{\dagger}$ of a continuous operator $\widehat{A} : \mathcal{S}(\mathbb{R}^n) \longrightarrow \mathcal{S}(\mathbb{R}^n)$ is defined by the formula

$$\langle \widehat{A}\psi | \phi \rangle = \langle \psi | \widehat{A}^{\dagger}\phi \rangle \tag{6.53}$$

for all $(\psi, \phi) \in \mathcal{S}(\mathbb{R}^n) \times \mathcal{S}(\mathbb{R}^n)$. Let us determine explicitly the Weyl symbol of $\widehat{A}^{\dagger}$.

Proposition 17 *The formal adjoint $\widehat{A}^{\dagger}$ of the Weyl operator $\widehat{A} = \mathrm{Op_W}(a)$ is the Weyl operator $\widehat{A}^{\dagger} = \mathrm{Op_W}(a^*)$. In particular, $\widehat{A}$ is formally self-adjoint if and only if a is a real function.*

Proof Expressing $\widehat{A}$ in terms of the Grossmann–Royer operators we have

$$\langle \widehat{A}\psi | \phi \rangle = \left(\tfrac{1}{\pi\hbar}\right)^n \int \left(\int a(z_0) \widehat{T}_{\mathrm{GR}}(z_0)\psi(x) d^{2n}z_0 \right)^* \phi(x) d^n x.$$

Let b be the Weyl symbol of $\widehat{A}^{\dagger}$; we have

$$(\widehat{A}^{\dagger}\psi | \phi) = \left(\tfrac{1}{\pi\hbar}\right)^n \int \left(\int b(z_0) \widehat{T}_{\mathrm{GR}}(z_0)\psi(x) d^{2n}z_0 \right)^* \phi(x) d^n x.$$

Since $\langle \widehat{A}^{\dagger}\psi | \phi \rangle = \langle \psi | \widehat{A}\phi \rangle$ for all $(\psi, \phi) \in \mathcal{S}(\mathbb{R}^n) \times \mathcal{S}(\mathbb{R}^n)$ we must have $a^* = b$. ■

The property that $\widehat{A}$ is formally self-adjoint if and only if a is a real function is very important in the applications to quantum mechanics, because the "quantization" of a real observable should precisely lead to a self-adjoint operator.

6.6.2 Composition Formulas

We now assume that the Weyl operators $\widehat{A} = \mathrm{Op}(a)$ and $\widehat{B} = \mathrm{Op}(b)$ can be composed and set $\widehat{C} = \widehat{A}\widehat{B}$. Assuming that we can write

$$\widehat{C} = \left(\tfrac{1}{2\pi\hbar}\right)^n \int c_\sigma(z) \widehat{T}(z) d^{2n}z$$

we ask: what is c_σ? The answer is given by the following result:

Proposition 18 *Let $\widehat{A} = \mathrm{Op}(a)$ and $\widehat{B} = \mathrm{Op}(b)$. (i) The product $\widehat{C} = \widehat{A}\widehat{B}$ has (when defined) Weyl symbol*

$$c(z) = \left(\tfrac{1}{4\pi\hbar}\right)^{2n} \int e^{\frac{i}{2\hbar}\sigma(z',z'')} a(z + \tfrac{1}{2}z') b(z - \tfrac{1}{2}z'') d^{2n}z' d^{2n}z''. \tag{6.54}$$

(ii) The covariant Weyl symbol of $\widehat{C}$ is given by

$$c_\sigma(z) = \left(\tfrac{1}{2\pi\hbar}\right)^n \int e^{\frac{i}{2\hbar}\sigma(z,z')} a_\sigma(z-z')b_\sigma(z')d^{2n}z'. \tag{6.55}$$

Proof Assume that the Weyl symbols a, b of $\widehat{A}$ and $\widehat{B}$ are in $\mathcal{S}(\mathbb{R}^{2n})$. We denote by $K_{\widehat{A}\widehat{B}}$ the distributional kernel of the product $\widehat{A}\widehat{B}$. We have

$$K_{\widehat{A}\widehat{B}}(x,y) = \left(\tfrac{1}{2\pi\hbar}\right)^{2n} \int e^{\frac{i}{\hbar}((x-\alpha)p+(\alpha-y)p)} \\ \times a(\tfrac{1}{2}(x+\alpha),\zeta)b(\tfrac{1}{2}(x+y),\xi)d^n\alpha d^n\zeta d^n\xi.$$

In view of formula (6.44) we have

$$c(x,p) = \int e^{-\frac{i}{\hbar}pu} K_{\widehat{A}\widehat{B}}(x+\tfrac{1}{2}u, x-\tfrac{1}{2}u)d^n u$$

and the symbol of $\widehat{A}\widehat{B}$ is thus

$$c(z) = \left(\tfrac{1}{\pi\hbar}\right)^{2n} \int e^{\frac{i}{\hbar}Q} a(\tfrac{1}{2}(x+\alpha+\tfrac{1}{2}u),\zeta) \\ \times b(\tfrac{1}{2}(x+\alpha-\tfrac{1}{2}u),\xi)d^n\alpha d^n\zeta d^n\xi$$

where the phase Q is given by

$$Q = (x-\alpha+\tfrac{1}{2}u)\zeta + (\alpha - x + \tfrac{1}{2}u)\xi - up \\ = (x-\alpha+\tfrac{1}{2}u)(\zeta - p) + (\alpha - x + \tfrac{1}{2}u)(\xi - p).$$

Setting $\zeta' = \zeta - p$, $\xi' = \xi - p$, $\alpha' = \tfrac{1}{2}(\alpha - x + \tfrac{1}{2}u)$ and $u' = \tfrac{1}{2}(\alpha - x - \tfrac{1}{2}u)$ we have

$$d^n\alpha d^n\zeta d^n u d^n\xi = 2^{2n} d^n\alpha' d^n\zeta' d^n u' d^n\xi'$$

and $Q = 2\sigma(u',\xi';\alpha',\zeta')$, hence

$$c(z) = \left(\tfrac{1}{\pi\hbar}\right)^{2n} \int e^{\frac{2i}{\hbar}\sigma(u',\xi';\alpha',\zeta')} a(x+\alpha', p+\zeta') \\ \times b(x+u', p+\xi')d^n\alpha' d^n\zeta' d^n u' d^n\xi';$$

formula (6.54) follows setting $z' = 2(\alpha',\zeta')$ and $z'' = -2(u',\xi')$. (ii) Writing the operators $\widehat{A}$ and $\widehat{B}$ in the form

$$\widehat{A} = \left(\tfrac{1}{2\pi\hbar}\right)^n \int a_\sigma(z_0)\widehat{T}(z_0)d^{2n}z_0$$

$$\widehat{B} = \left(\tfrac{1}{2\pi\hbar}\right)^n \int b_\sigma(z_1)\widehat{T}(z_1)d^{2n}z_1$$

we have, using the property (6.15) of the Heisenberg operators

$$\begin{aligned}\widehat{T}(z_0)\widehat{B} &= \left(\tfrac{1}{2\pi\hbar}\right)^n \int b_\sigma(z_1)\widehat{T}(z_0)\widehat{T}(z_1)d^{2n}z_1 \\ &= \left(\tfrac{1}{2\pi\hbar}\right)^n \int e^{\frac{i}{2\hbar}\sigma(z_0,z_1)}b_\sigma(z_1)\widehat{T}(z_0+z_1)d^{2n}z_1\end{aligned}$$

and hence

$$\widehat{A}\widehat{B} = \left(\tfrac{1}{2\pi\hbar}\right)^{2n} \int e^{\frac{i}{2\hbar}\sigma(z_0,z_1)}a_\sigma(z_0)b_\sigma(z_1)\widehat{T}(z_0+z_1)d^{2n}z_0d^{2n}z_1.$$

Setting $z = z_0 + z_1$ and $z' = z_1$ this can be written

$$\widehat{A}\widehat{B} = \left(\tfrac{1}{2\pi\hbar}\right)^{2n} \int \left(\int e^{\frac{i}{2\hbar}\sigma(z,z')}a_\sigma(z-z')b_\sigma(z')d^{2n}z'\right)\widehat{T}(z)d^{2n}z$$

hence (6.55). ∎

The function c defined by (6.54) is often called the Moyal (or Groenewold–Moyal) starproduct, and is denoted by $a \star_\hbar b$.

References

1. J. Chazarain, A. Piriou, *Introduction à la théorie des équations aux dérivées partielles linéaires* (Gauthier-Villars, Paris, 1981). English translation: *Introduction to the theory of linear partial differential equations. Studies in Mathematics and its Applications*, vol. 14 (North-Holland, 1982)
2. L. Cohen, *The Weyl Operator and Its Generalization* (Springer Science & Business Media, 2012)
3. H.O. Cordes, On compactness of commutators of multiplications and convolutions, and boundedness of pseudodifferential symbols, J. Funct. Anal. **18**, 115–131 (1975)
4. M. de Gosson, *Symplectic Geometry and Quantum Mechanics*. Operator Theory: Advances and Applications. Advances in Partial Differential Equations, vol. 166 (Birkhäuser, Basel, 2006)
5. M. de Gosson, *Symplectic Methods in Harmonic Analysis and in Mathematical Physics* (Birkhäuser, 2011)
6. G.B. Folland, *Harmonic Analysis in Phase Space. Annals of Mathematics Studies* (Princeton University Press, Princeton, 1989)
7. A. Grossmann, Parity operators and quantization of δ-functions. Commun. Math. Phys. **48**, 191–193 (1976)
8. L. Hörmander, *The Analysis of Linear Partial Differential Operators III* (Springer, Berlin, 1985)

9. R.G. Littlejohn, The semiclassical evolution of wave packets. Phys. Rep. **138**(4–5), 193–291 (1986)
10. A. Royer, Wigner functions as the expectation value of a parity operator. Phys. Rev. A **15**, 449–450 (1977)
11. M.A. Shubin, *Pseudodifferential Operators and Spectral Theory* (Springer-Verlag, 1987) (original Russian edition in Nauka, Moskva (1978))
12. H. Weyl, Quantenmechanik und Gruppentheorie. Zeitschrift für Physik, 46 (1927)
13. H. Weyl, *The Theory of Groups and Quantum Mechanics*, translated from the 2nd German edition by H.P. Robertson (New York, Dutten, 1931)
14. M.W. Wong, *Weyl Transforms* (Springer, 1998)

Chapter 7
The Cohen Class

In this chapter we study the "Cohen class", which is a family of particular phase space quasi-distributions characterized by two simple properties (continuity, and translation covariance). These quasi-distributions are obtained from the Wigner transform by convolving the latter with a suitable tempered distribution. The Cohen class is widely used in time-frequency analysis and is being rediscovered in quantum mechanics.

7.1 The Wigner and Ambiguity Transforms

Recall (Examples 2 and 3 of the last chapter) that the Wigner and ambiguity transforms of a function were defined by

$$\operatorname{Wig}\psi(z) = \left(\tfrac{1}{\pi\hbar}\right)^n \langle \widehat{T}_{\mathrm{GR}}(z)\psi|\psi\rangle \tag{7.1}$$

$$\operatorname{Amb}\psi(z) = \left(\tfrac{1}{2\pi\hbar}\right)^n \langle \widehat{T}(z)\psi|\psi\rangle \tag{7.2}$$

where

$$\begin{aligned}\widehat{T}_{\mathrm{GR}}(z_0)\psi(x) &= e^{\frac{2i}{\hbar}p_0(x-x_0)}\psi(2x_0-x)\\ \widehat{T}(z_0)\psi(x) &= e^{\frac{i}{\hbar}(p_0x-\frac{1}{2}p_0x_0)}\psi(x-x_0).\end{aligned}$$

are, respectively the Grossmann–Royer and Heisenberg operators. In this section we extend these definitions to *pairs* of functions.

7.1.1 The Cross-Wigner Transform

It is convenient, having applications to Weyl calculus in mind, to give the following non-standard definition of the (cross-)Wigner transform:

M.A. de Gosson, *Born–Jordan Quantization*, Fundamental Theories of Physics 182, DOI 10.1007/978-3-319-27902-2_7

Definition 1 Let $(\psi, \phi) \in \mathcal{S}(\mathbb{R}^n) \times \mathcal{S}(\mathbb{R}^n)$. The function $(\psi, \phi) \longmapsto \mathrm{Wig}(\psi, \phi)$ defined by

$$\mathrm{Wig}(\psi, \phi)(z) = \left(\tfrac{1}{\pi\hbar}\right)^n \langle \widehat{T}_{\mathrm{GR}}(z)\phi|\psi\rangle \tag{7.3}$$

is called the cross-Wigner function (or transform) of (ψ, ϕ).

When $\psi = \phi$ we recover the usual Wigner distribution $\mathrm{Wig}\,\psi$.

The physical interpretation of the cross-Wigner transform is that of an interference term in the Wigner distribution of the sum $\mathrm{Wig}(\psi, \phi)$:

$$\mathrm{Wig}(\psi + \phi) = \mathrm{Wig}\,\phi + \mathrm{Wig}\,\phi + 2\,\mathrm{Re}\,\mathrm{Wig}(\psi, \phi).$$

We mention that the importance of these interference terms has been emphasized and studied by Zurek [10] in the context of the sub-Planckian structures in phase space. They also appear in the study of weak values, as we will see in Chap. 11.

Using the explicit definition of the Grossmann–Royer operator one immediately recovers the text-book formulas [3–6]

$$\mathrm{Wig}(\psi, \phi)(z) = \left(\tfrac{1}{2\pi\hbar}\right)^n \int e^{-\frac{i}{\hbar}py}\psi(x + \tfrac{1}{2}y)\phi^*(x - \tfrac{1}{2}y)d^n y \tag{7.4}$$

and

$$\mathrm{Wig}\,\psi(z) = \left(\tfrac{1}{2\pi\hbar}\right)^n \int e^{-\frac{i}{\hbar}py}\psi(x + \tfrac{1}{2}y)\psi^*(x - \tfrac{1}{2}y)d^n y \tag{7.5}$$

which are often taken as *definitions*. It turns out that $(\psi, \phi) \longmapsto \mathrm{Wig}(\psi, \phi)$ is a sesquilinear mapping $L^2(\mathbb{R}^n) \times L^2(\mathbb{R}^n) \longrightarrow L^2(\mathbb{R}^{2n})$ (see Proposition 6); i.e.

$$\mathrm{Wig}(\psi_1 + \psi_2, \phi) = \mathrm{Wig}(\psi_1, \phi) + \mathrm{Wig}(\psi_2, \phi)$$
$$\mathrm{Wig}(\psi, \phi_1 + \phi_2) = \mathrm{Wig}(\psi, \phi_1) + \mathrm{Wig}(\psi, \phi_2)$$

and

$$\mathrm{Wig}(\lambda\psi, \phi) = \lambda \mathrm{Wig}(\psi, \phi),\ \mathrm{Wig}(\psi, \lambda\phi) = \lambda^* \mathrm{Wig}(\psi, \phi)$$

for every complex number[1] λ. The following complex conjugation property is also obvious:

$$\mathrm{Wig}(\psi, \phi) = \mathrm{Wig}(\phi, \psi)^* \tag{7.6}$$

and hence, in particular, $\mathrm{Wig}\,\psi$ is always a real function. Notice that $\mathrm{Wig}\,\psi$ however in general takes negative values. In fact, we have the following well-known result,

---

[1] In some physical texts the definition of the cross-Wigner transform is the complex conjugate of ours. This leads to sesquilinearity properties which are in accordance with that of the inner product $\langle\cdot|\cdot\rangle$.

which was proven by Hudson [7] in the case $n = 1$, and in the general case by Janssen [8]: we have $\operatorname{Wig}\psi \geq 0$ if and only if ψ is a generalized Gaussian, that is, a function of the type

$$\psi(x) = Ce^{-M(x-x_0)^2}$$

where C is a complex coefficient and M a symmetric $n \times n$ matrix such that $\operatorname{Re} M > 0$.

Example 2 Let $\phi_0(x) = (\pi\hbar)^{-n/4}e^{-|x|^2/2\hbar}$ be the standard coherent state. We have

$$\operatorname{Wig}\phi_0(x,p) = (\pi\hbar)^{-n}e^{-\frac{1}{\hbar}(|x|^2+|p|^2)}.$$

One shows that the mapping $(\psi, \phi) \longmapsto \operatorname{Wig}(\psi, \phi)$, which is continuous from $\mathcal{S}(\mathbb{R}^n) \times \mathcal{S}(\mathbb{R}^n)$ to $\mathcal{S}(\mathbb{R}^{2n})$, extends to a continuous mapping

$$\operatorname{Wig} : \mathcal{S}'(\mathbb{R}^n) \times \mathcal{S}(\mathbb{R}^n) \longrightarrow \mathcal{S}'(\mathbb{R}^{2n})$$

(this property, which is certainly not obvious if one uses the traditional definition (7.4) is easy to prove if one rewrites the definition (7.3) using distributional brackets $\langle \cdot, \cdot \rangle$ in place of the L^2 scalar product:

$$\operatorname{Wig}(\psi, \phi)(z) = \left(\tfrac{1}{\pi\hbar}\right)^n \langle (\widehat{T}_{\mathrm{GR}}(z)\phi)^*, \psi\rangle.$$

Similarly, using the fact that the map $z \longmapsto \widehat{T}_{\mathrm{GR}}(z)$ is strongly continuous on $\mathcal{S}(\mathbb{R}^n)$ and weakly $*$-continuous on $\mathcal{S}'(\mathbb{R}^n)$, one proves that if $\psi \in \mathcal{S}'(\mathbb{R}^n)$ and $\phi \in \mathcal{S}(\mathbb{R}^n)$ then $\operatorname{Wig}(\psi, \phi)$ is continuous on $\mathbb{R}^{2n}$ (see de Gosson [4, 5]).

The physical interest of the Wigner transform comes from the following result: for every $\psi \in L^1(\mathbb{R}^n) \cap L^2(\mathbb{R}^n)$ such that $\widehat{\psi} \in L^1(\mathbb{R}^n)$ ($\widehat{\psi} = F\psi$, the Fourier transform of ψ), we have

$$\int \operatorname{Wig}\psi(x,p)d^np = |\psi(x)|^2, \int \operatorname{Wig}\psi(x,p)d^nx = |\widehat{\psi}(p)|^2 \qquad (7.7)$$

(see for instance [3, 5]). In particular if $\int \operatorname{Wig}\psi(z)d^{2n}z = 1$ then $\operatorname{Wig}\psi$ can be interpreted as a quasi-distribution (in the probabilistic sense) since the formulas above imply that $||\psi|| = ||\widehat{\psi}|| = 1$ so that $|\psi|^2$ and $|\widehat{\psi}|^2$ then play the role of marginal (quasi-)probability densities. More generally:

Proposition 3 *Let* $(\psi, \phi) \in \mathcal{S}(\mathbb{R}^n) \times \mathcal{S}(\mathbb{R}^n)$. *We have*

$$\int \operatorname{Wig}(\psi, \phi)(z)d^np = \psi(x)\phi^*(x) \qquad (7.8)$$

$$\int \operatorname{Wig}(\psi, \phi)(z)d^nx = \widehat{\psi}(p)\widehat{\phi}^*(p) \qquad (7.9)$$

where $\widehat{\psi} = F\psi$ *is the* $\hbar$*-Fourier transform of* ψ. *Hence, in particular*

$$\int \mathrm{Wig}(\psi,\phi)(z)d^{2n}z = \langle\phi|\psi\rangle. \tag{7.10}$$

Proof Formula (7.10) immediately follows from any of the two formulas (7.8) and (7.9). Let us prove (7.8). We have

$$\begin{aligned}
\int \mathrm{Wig}(\psi,\phi)(z)d^{n}p &= \left(\tfrac{1}{2\pi\hbar}\right)^{n} \int \left(\int e^{-\frac{i}{\hbar}py}\psi(x+\tfrac{1}{2}y)\phi^{*}(x-\tfrac{1}{2}y)d^{n}y \right) d^{n}p \\
&= \left(\tfrac{1}{2\pi\hbar}\right)^{n} \int \left(\int e^{-\frac{i}{\hbar}py}d^{n}p \right) \psi(x+\tfrac{1}{2}y)\phi^{*}(x-\tfrac{1}{2}y)d^{n}y \\
&= \int \delta(y)\psi(x+\tfrac{1}{2}y)\phi^{*}(x-\tfrac{1}{2}y)d^{n}y \\
&= \left(\int \delta(y)d^{n}y \right) \psi(x)\phi^{*}(x)
\end{aligned}$$

hence formula (7.8). To prove (7.9) we proceed as follows: writing

$$\begin{aligned}
\int \mathrm{Wig}(\psi,\phi)(z)d^{n}x &= \left(\tfrac{1}{2\pi\hbar}\right)^{n} \int \left(\int e^{-\frac{i}{\hbar}py}\psi(x+\tfrac{1}{2}y)\phi^{*}(x-\tfrac{1}{2}y)d^{n}y \right) d^{n}x \\
&= \left(\tfrac{1}{2\pi\hbar}\right)^{n} \int e^{-\frac{i}{\hbar}py} \left(\int \psi(x+\tfrac{1}{2}y)\phi^{*}(x-\tfrac{1}{2}y)d^{n}x \right) d^{n}y
\end{aligned}$$

we notice that the integral between the brackets is a convolution product:

$$\int \psi(x+\tfrac{1}{2}y)\phi^{*}(x-\tfrac{1}{2}y)d^{n}x = \psi * (\phi^{*})^{\vee}$$

(here $^{\vee}$ is the reflection operator $x \longmapsto -x$); formula (7.9) now follows using the formulas

$$F(\psi * (\phi^{*})^{\vee}) = (2\pi\hbar)^{n} F\psi F\left[(\phi^{*})^{\vee}\right]$$

and $F\left[(\phi^{*})^{\vee}\right] = (F\phi^{*})$. ■

The result above holds for less stringent conditions on the functions ψ and ϕ; for instance it suffices that $\psi,\phi \in L^{1}(\mathbb{R}^{n}) \cap L^{2}(\mathbb{R}^{n})$ as closer scrutiny of the proof shows.

7.1.2 The Cross-Ambiguity Function

We have defined the cross-Wigner transform using the Grossmann-Royer operator; if one uses the Heisenberg operator one obtains the cross-ambiguity transform, familiar from radar theory (see Folland [3], Gröchenig [6]):

Definition 4 Let $(\psi, \phi) \in \mathcal{S}(\mathbb{R}^n) \times \mathcal{S}(\mathbb{R}^n)$. The function $(\psi, \phi) \longmapsto \operatorname{Amb}(\psi, \phi)$ defined by

$$\operatorname{Amb}(\psi, \phi)(z) = \left(\tfrac{1}{2\pi\hbar}\right)^n \langle \widehat{T}(z)\phi|\psi\rangle \tag{7.11}$$

is called the cross-ambiguity function (or transform) of (ψ, ϕ). The function

$$\operatorname{Amb}\psi = \operatorname{Amb}(\psi, \psi)$$

is called the ambiguity function of ψ.

Recalling that the Heisenberg operator is defined by

$$\widehat{T}(z_0)\psi(x) = e^{\frac{i}{\hbar}(p_0x - \frac{1}{2}p_0x_0)}\psi(x - x_0)$$

it follows that we have the explicit expression

$$\operatorname{Amb}(\psi, \phi)(z) = \left(\tfrac{1}{2\pi\hbar}\right)^n \int e^{-\frac{i}{\hbar}py}\psi(y + \tfrac{1}{2}x)\phi^*(y - \tfrac{1}{2}x)d^ny. \tag{7.12}$$

The cross-Wigner transform satisfies the conjugation relation $\operatorname{Wig}(\psi, \phi) = \operatorname{Wig}(\phi, \psi)^*$ (formula (7.6)); this relation is not preserved for the cross-ambiguity function; instead we have

$$\operatorname{Amb}(\psi, \phi)^* = \operatorname{Amb}(\phi^\vee, \psi^\vee) \tag{7.13}$$

as is easily verified using for instance (7.12); here $\psi^\vee(x) = \psi(-x)$.

The fundamental relation between the cross-Wigner and the cross-ambiguity transforms can be proven using formula (6.26) relating the Heisenberg and the Grossmann–Royer operators (cf. Corollary 6 in Chap. 6); we give here a direct proof.

Proposition 5 *Let $(\psi, \phi) \in \mathcal{S}(\mathbb{R}^n) \times \mathcal{S}(\mathbb{R}^n)$. We have*

$$\operatorname{Amb}(\psi, \phi) = F_\sigma \operatorname{Wig}(\psi, \phi) \tag{7.14}$$

$$\operatorname{Wig}(\psi, \phi) = F_\sigma \operatorname{Amb}(\psi, \phi). \tag{7.15}$$

Proof (This proof corrects the proof of Proposition 175(i) in [2], §9.3.1). It is sufficient to prove formula (7.14) since F_σ is involutive. Set

$$A(z) = (2\pi\hbar)^{2n} F_\sigma \operatorname{Wig}(\psi, \phi)(z);$$

by definition of F_σ and $\operatorname{Wig}(\psi, \phi)$ we have

$$A(z) = \int e^{-\frac{i}{\hbar}\sigma(z,z')} \left(\int e^{-\frac{i}{\hbar}p'y}\psi(x' + \tfrac{1}{2}y)\phi^*(x' - \tfrac{1}{2}y)d^ny \right) d^{2n}z'$$

$$= \int e^{-\frac{i}{\hbar}p'(y-x)} e^{-\frac{i}{\hbar}px'} \psi(x' + \tfrac{1}{2}y)\phi^*(x' - \tfrac{1}{2}y) d^n p' d^n x' d^n y$$

$$= \int \left(\int e^{-\frac{i}{\hbar}p'(y-x)} d^n p' \right) e^{-\frac{i}{\hbar}px'} \psi(x' + \tfrac{1}{2}y)\phi^*(x' - \tfrac{1}{2}y) d^n x' d^n y.$$

Using the Fourier transform formula

$$\int e^{-\frac{i}{\hbar}p'(y-x)} d^n p' = (2\pi\hbar)^n \, \delta(x - y)$$

we can rewrite $A(z)$ as

$$A(z) = (2\pi\hbar)^n \int e^{-\frac{i}{\hbar}px'} \delta(x - y)\psi(x' + \tfrac{1}{2}y)\phi^*(x' - \tfrac{1}{2}y) d^n x' d^n y$$

$$= (2\pi\hbar)^n \int e^{-\frac{i}{\hbar}px'} \psi(x' + \tfrac{1}{2}x)\phi^*(x' - \tfrac{1}{2}x) d^n x'$$

hence $F_\sigma \mathrm{Wig}(\psi, \phi) = \mathrm{Amb}(\psi, \phi)$ as claimed. ■

The cross-Wigner and cross-ambiguity functions both satisfy the important "Moyal identities":

Proposition 6 *Denote by* $\langle\langle\cdot|\cdot\rangle\rangle$ *the* L^2*-scalar product on* $\mathbb{R}^{2n}$*. (i) For all* ψ, ψ', ϕ, ϕ' *in* $L^2(\mathbb{R}^n) \times L^2(\mathbb{R}^n)$ *we have*

$$\langle\langle \mathrm{Wig}(\psi, \phi)|\mathrm{Wig}(\psi', \phi')\rangle\rangle = \left(\tfrac{1}{2\pi\hbar}\right)^n \langle\psi|\psi'\rangle\langle\phi|\phi'\rangle^* \tag{7.16}$$

and

$$\langle\langle \mathrm{Amb}(\psi, \phi)|\mathrm{Amb}(\psi', \phi')\rangle\rangle = \left(\tfrac{1}{2\pi\hbar}\right)^n \langle\psi|\psi'\rangle\langle\phi|\phi'\rangle^*. \tag{7.17}$$

(ii) We have $\mathrm{Wig}\,\psi \in L^2(\mathbb{R}^{2n})$ *(resp.* $\mathrm{Amb}\,\psi \in L^2(\mathbb{R}^{2n})$*) if and only if* $\psi \in L^2(\mathbb{R}^n)$

$$|||\mathrm{Wig}\,\psi||| = |||\mathrm{Amb}\,\psi||| = \left(\tfrac{1}{2\pi\hbar}\right)^{n/2} ||\psi||$$

where $|||\cdot|||$ *is the* L^2*-norm on* $\mathbb{R}^{2n}$*. In particular,* Wig *and* Amb *are continuous mappings* $L^2(\mathbb{R}^n) \times L^2(\mathbb{R}^n) \longrightarrow L^2(\mathbb{R}^{2n})$.

Proof Property (ii) immediately follows from (i) taking $\psi = \psi' = \phi = \phi'$ in (7.16) and (7.17), respectively. (i) Each of the two formulas (7.16) and (7.17) is deduced from the other using Plancherel's formula (6.22) for symplectic Fourier transforms. Let us prove (7.16). We begin by noting that the scalar product

$$A = (2\pi\hbar)^{2n} \langle\langle \mathrm{Wig}(\psi, \phi)|\mathrm{Wig}(\psi', \phi')\rangle\rangle$$

is given by the expression

$$\int e^{-\frac{i}{\hbar}p(y'-y)}\psi^*(x+\tfrac{1}{2}y)\psi'(x+\tfrac{1}{2}y')$$
$$\times\,\phi(x-\tfrac{1}{2}y)\phi'^*(x-\tfrac{1}{2}y')d^ny d^ny' d^nx d^np.$$

The integral in p (interpreted as a distributional bracket) is

$$\int e^{-\frac{i}{\hbar}p(y'-y)}d^np = (2\pi\hbar)^n\delta(y-y')$$

and hence

$$A = (2\pi\hbar)^n\int \psi^*(x+\tfrac{1}{2}y)\psi'(x-\tfrac{1}{2}y)$$
$$\times\,\phi(x+\tfrac{1}{2}y)\phi'^*(x-\tfrac{1}{2}y)d^ny d^ny' d^nx.$$

Setting $u = x+\frac{1}{2}y$ and $v = x-\frac{1}{2}y$ we have $d^nud^nv = d^nxd^ny$ hence

$$\begin{aligned} A &= (2\pi\hbar)^n\left(\int \psi^*(u)\psi'(u)du\right)\left(\int \phi(v)\phi'^*(v)dv\right) \\ &= \langle\psi|\psi'\rangle\langle\phi|\phi'\rangle^* \end{aligned}$$

proving the identity (7.16). ∎

The Moyal identities (and their extensions to more general quasi-distributions) play a very important role in many topics in harmonic analysis. We are going to see below that it allows to prove a reconstruction formula, which we will later use when we will deal with the notion of weak value in Chap. 11.

7.1.3 A Reconstruction Formula

It is easy to see using the properties of the Fourier transform that the datum of the Wigner transform $\mathrm{Wig}\,\psi$ determines ψ up to a constant factor with modulus one (it thus determines unambiguously the quantum state $|\psi\rangle$. In the case of the cross-Wigner transform, one can prove an interesting inversion formula:

Proposition 7 *Let $\phi,\gamma \in L^2(\mathbb{R}^n)$ be non-orthogonal: $\langle\phi|\gamma\rangle \neq 0$. For every $\psi \in S'(\mathbb{R}^n)$ we have*

$$\psi(x) = \frac{2^n}{\langle\phi|\gamma\rangle}\int \mathrm{Wig}(\psi,\phi)(z_0)\widehat{T}_{\mathrm{GR}}(z_0)\gamma(x)d^{2n}z_0. \tag{7.18}$$

Proof Let us denote by $\chi(x)$ the right hand side of (7.18):

$$\chi(x) = \frac{2^n}{\langle \phi|\gamma \rangle} \int \operatorname{Wig}(\psi, \phi)(z_0)\widehat{T}_{\mathrm{GR}}(z_0)\gamma(x)d^{2n}z_0.$$

This function is well-defined since $\operatorname{Wig}(\psi, \phi) \in L^2(\mathbb{R}^{2n})$ in view of Moyal's identity (7.16). For every $\alpha \in \mathcal{S}(\mathbb{R}^n)$ we have

$$\langle \alpha|\chi \rangle = \frac{2^n}{\langle \phi|\gamma \rangle} \int \operatorname{Wig}(\psi, \phi)(z)\langle \alpha|\widehat{T}_{\mathrm{GR}}(z)\gamma \rangle d^{2n}z.$$

Taking into account the formula

$$\operatorname{Wig}(\alpha, \gamma)(z) = \left(\tfrac{1}{\pi\hbar}\right)^n \langle \widehat{T}_{\mathrm{GR}}(z)\gamma|\alpha \rangle$$

and the conjugation relation $\operatorname{Wig}(\gamma, \alpha) = \operatorname{Wig}(\alpha, \gamma)^*$ we thus have

$$\begin{aligned}\langle \alpha|\chi \rangle &= \frac{(2\pi\hbar)^n}{\langle \phi|\gamma \rangle} \int \operatorname{Wig}(\psi, \phi)(z)W(\alpha, \gamma)^*(z)d^{2n}z \\ &= \frac{(2\pi\hbar)^n}{\langle \phi|\gamma \rangle}\langle\langle \operatorname{Wig}(\alpha, \gamma)|\operatorname{Wig}(\psi, \phi)\rangle\rangle\end{aligned}$$

where $\langle\langle \cdot|\cdot \rangle\rangle$ is the scalar product on $L^2(\mathbb{R}^n)$. Using Moyal's identity we get

$$\langle \alpha|\chi \rangle = \left(\frac{1}{2\pi\hbar}\right)^n \frac{(2\pi\hbar)^n}{\langle \phi|\gamma \rangle}\langle \alpha|\psi \rangle\langle \phi|\gamma \rangle = \langle \alpha|\psi \rangle.$$

Since this identity holds for all $\alpha \in \mathcal{S}(\mathbb{R}^n)$ we have $\chi = \psi$ almost everywhere, which proves formula (7.18). ■

This formula will be generalized later on when we study the notion of weak value of an observable.

7.1.4 Relation with the Weyl Correspondence

The relation between the cross-Wigner and ambiguity transforms and the Weyl correspondence comes from the following extremely important result, which is the link between the Weyl correspondence and the phase space formalism of quantum mechanics. We denote as usual by $\langle \cdot, \cdot \rangle$ and $\langle\langle \cdot, \cdot \rangle\rangle$ the distributional brackets on $\mathbb{R}^n$ and $\mathbb{R}^{2n}$, respectively.

Proposition 8 *Let $a \in \mathcal{S}'(\mathbb{R}^{2n})$ and $\widehat{A} = \operatorname{Op_W}(a)$. (i) We have*

$$\langle \widehat{A}\psi, \phi^* \rangle = \langle\langle a, \operatorname{Wig}(\psi, \phi)\rangle\rangle \tag{7.19}$$

for all $(\psi, \phi) \in \mathcal{S}(\mathbb{R}^n) \times \mathcal{S}(\mathbb{R}^n)$. *In integral notation:*

$$\int \widehat{A}\psi(x)\phi^*(x)d^n x = \int a(z)\mathrm{Wig}(\psi, \phi)(z)d^{2n}z. \tag{7.20}$$

(ii) Similarly,

$$\int \widehat{A}\psi(x)\phi^*(x)d^n x = \int a_\sigma(z)\mathrm{Amb}(\psi, \phi)(-z)d^{2n}z. \tag{7.21}$$

Proof (i) In view of definition (6.36) we have, using Fubini's theorem and definition (7.3) of the Wigner function,

$$\begin{aligned}
\langle \widehat{A}\psi, \phi^* \rangle &= \left(\tfrac{1}{\pi\hbar}\right)^n \int \left(\int a(z_0)\widehat{T}_{\mathrm{GR}}(z_0)\psi(x)d^{2n}z_0 \right) \phi^*(x)d^n x \\
&= \left(\tfrac{1}{\pi\hbar}\right)^n \int a(z_0) \left(\int \widehat{T}_{\mathrm{GR}}(z_0)\psi(x)\phi^*(x)d^n x \right) d^{2n}z_0 \\
&= \int a(z_0)\mathrm{Wig}(\psi, \phi)(z_0)d^{2n}z_0
\end{aligned}$$

which is (7.19). (ii) Using Plancherel's formula (6.23) we have

$$\int a(z_0)\mathrm{Wig}(\psi, \phi)(z_0)d^{2n}z_0 = \int F_\sigma a(z_0)F_\sigma \mathrm{Wig}(\psi, \phi)(-z_0)d^{2n}z_0$$

which implies (7.21). ∎

Formula (7.19) can be taken as a *definition* of the Weyl operator $\widehat{A} = \mathrm{Op_W}(a)$: it is the only operator $\mathcal{S}(\mathbb{R}^n) \longrightarrow \mathcal{S}'(\mathbb{R}^n)$ for which the equality (7.19) holds. We will exploit this fact when we define the Born–Jordan operators.

When $\widehat{A}$ is self-adjoint on $L^2(\mathbb{R}^n)$ we recover, taking $\psi = \phi$, the following familiar result from quantum mechanics:

$$\langle \psi|\widehat{A}|\psi \rangle = \int a(z)\mathrm{Wig}\,\psi(z)d^{2n}z \tag{7.22}$$

which clearly shows how the Wigner transform indeed plays the role of a probability quasi-distribution: if ψ is normalized to unity then $\langle \widehat{A} \rangle_\psi = \langle \psi|\widehat{A}|\psi \rangle$ is the mean value of $\widehat{A}$ in the pure state $|\psi\rangle$ while the integral of $\mathrm{Wig}\,\psi$ over $\mathbb{R}^{2n}$ is equal to one—as would be the case for a true probability distribution.

7.2 The Cohen Class

7.2.1 Definition and General Properties

The lack of positivity of the Wigner distribution $\mathrm{Wig}\,\psi$ which makes its interpretation as a true probability density problematic has led to a search for alternative distributions $Q\psi$. One of the most famous examples is Husimi's distribution, which is the convolution of the Wigner transform with a Gaussian function. More generally, we will say following Gröchenig [6], §4.5, that:

Definition 9 Let $Q : \mathcal{S}(\mathbb{R}^n) \times \mathcal{S}(\mathbb{R}^n) \longrightarrow \mathcal{S}'(\mathbb{R}^{2n})$ be a sesquilinear form. We say that Q belongs to the *Cohen class* if we have

$$Q(\psi, \phi) = \mathrm{Wig}\,(\psi, \phi) * \theta \tag{7.23}$$

for some distribution $\theta \in \mathcal{S}'(\mathbb{R}^{2n})$. The function θ is called a Cohen kernel.

The cross-Wigner transform trivially belongs to the Cohen class (take $\theta = \delta$, the Dirac distribution on $\mathbb{R}^{2n}$).

The Husimi distribution is a typical (and well studied) member of the Cohen class:

Example 10 The member of the Cohen class corresponding to the of choice of the normalized phase space Gaussian

$$\theta(z) = (\pi\hbar)^{-n} e^{-\frac{1}{\hbar}|z|^2}$$

is called the Husimi distribution. Since $\theta(z)$ is the Wigner transform of the standard coherent state $\phi_0(x) = (\pi\hbar)^{-n/4} e^{-|x|^2/2\hbar}$ (Example 2) the Husimi distribution is always positive.

The following result gives a sufficient (but not necessary) condition for a sesquilinear form to belong to Cohen's class.

Proposition 11 *Let $Q : \mathcal{S}(\mathbb{R}^n) \times \mathcal{S}(\mathbb{R}^n) \longrightarrow \mathcal{S}(\mathbb{R}^{2n})$ be a sesquilinear form and set $Q\psi = Q(\psi, \psi)$. If Q is such that*

$$Q\psi(z - z_0) = Q(\widehat{T}(z_0)\psi)(z) \tag{7.24}$$

$$|Q(\psi, \phi)(0, 0)| \leq ||\psi||\,||\phi|| \tag{7.25}$$

for all ψ, ϕ in $L^2(\mathbb{R}^n)$ then there exists a distribution $\theta \in \mathcal{S}'(\mathbb{R}^{2n})$ such that $Q\psi = \mathrm{Wig}\,\psi * \theta$ for all $\psi \in \mathcal{S}(\mathbb{R}^n)$.

Proof (Gröchenig [6] proves this for $\hbar = 1/2\pi$; we are following here de Gosson [4]). The condition (7.25) means that the function $(\psi, \phi) \longmapsto Q(\psi, \phi)(0, 0)$ is a bounded sesquilinear form. Hence, by Riesz's representation theorem, there exists a bounded operator $\widehat{A}$ on $L^2(\mathbb{R}^n)$ such that

$$Q(\psi, \phi)(0,0) = \langle \phi | \widehat{A}\psi \rangle.$$

In view of (7.24) we have

$$\begin{aligned} Q\psi(z_0) &= Q(\widehat{T}(-z_0)\psi)(0) \\ &= \langle \widehat{A}\widehat{T}(-z_0)\psi, \widehat{T}(-z_0)\psi \rangle. \end{aligned}$$

In view of Schwartz's kernel theorem there exists a distribution $K \in \mathcal{S}'(\mathbb{R}^n \times \mathbb{R}^n)$ such that

$$\langle \phi | \widehat{A}\psi \rangle = \langle\langle K, \psi \otimes \phi^* \rangle\rangle$$

for all $\psi, \phi \in \mathcal{S}(\mathbb{R}^n)$ ($\langle\langle \cdot, \cdot \rangle\rangle$ is the distributional bracket on $\mathbb{R}^{2n}$). We thus have

$$\begin{aligned} Q\psi(z_0) &= \langle\langle K, \widehat{T}(-z_0)\psi \otimes (\widehat{T}(-z_0)\psi)^* \rangle\rangle \\ &= \int K(x,y)\widehat{T}(-z_0)\psi(x)(\widehat{T}(-z_0)\psi(y))^* d^n x d^n y. \end{aligned}$$

By definition of the Weyl–Heisenberg operators we have

$$\begin{aligned} \widehat{T}(-z_0)\psi(x) &= e^{\frac{i}{\hbar}(-p_0 x - \frac{1}{2}p_0 x_0)}\psi(x + x_0) \\ (\widehat{T}(-z_0)\psi(y))^* &= e^{-\frac{i}{\hbar}(-p_0 y - \frac{1}{2}p_0 x_0)}\psi^*(y + x_0) \end{aligned}$$

and hence

$$Q\psi(z_0) = \int e^{-\frac{i}{\hbar}p_0(x-y)} K(x,y)\psi(x+x_0)\psi^*(y+x_0)d^n x d^n y. \tag{7.26}$$

On the other hand, for every $\theta \in \mathcal{S}'(\mathbb{R}^{2n})$ we have

$$(\operatorname{Wig}\psi * \theta)(z_0) = \int \operatorname{Wig}\psi(z_0 - z)\theta(z)d^{2n}z$$

(the integral being interpreted in the distributional sense) hence, in view of the definition of the Wigner transform,

$$\begin{aligned} (\operatorname{Wig}\psi * \theta)(z_0) = \left(\tfrac{1}{2\pi\hbar}\right)^n \int & e^{-\frac{i}{\hbar}(p_0 - p)y'} \psi(x_0 - x' + \tfrac{1}{2}y') \\ & \times \psi^*(x_0 - x' - \tfrac{1}{2}y')\theta(x', p')d^n p d^n x' d^n y' \end{aligned}$$

that is, calculating the integral in the p variables,

$$\begin{aligned} (\operatorname{Wig}\psi * \theta)(z_0) = \left(\tfrac{1}{2\pi\hbar}\right)^{n/2} \int & F_2^{-1}\theta(x', y')e^{-\frac{i}{\hbar}p_0 y'} \psi(x_0 - x' + \tfrac{1}{2}y') \\ & \times \psi^*(x_0 - x' - \tfrac{1}{2}y')\theta(x', p')d^n x' d^n y' \end{aligned}$$

where $F_2^{-1}\theta$ is the inverse Fourier transform of θ in the second set of variables. Making the change of variables $x' = -\frac{1}{2}(x+y)$, $y' = x - y$ we have $d^n x' d^n y' = d^n x d^n y$ and the equality above becomes

$$(\operatorname{Wig}\psi * \theta)(z_0) = \left(\tfrac{1}{2\pi\hbar}\right)^{n/2} \int F_2^{-1}\theta(x, x-y)$$
$$\times e^{-\frac{i}{\hbar}p_0(x-y)}\psi(x+x_0)\psi^*(y+x_0)d^n x d^n y.$$

Comparing with formulas (7.26) we see that $Q\psi = \operatorname{Wig}\psi * \theta$ where θ is given by

$$K(x,y) = \left(\tfrac{1}{2\pi\hbar}\right)^{n/2} F_2^{-1}\theta(x, x-y)$$

that is

$$\theta(x,p) = (2\pi\hbar)^{n/2} \int e^{-\frac{i}{\hbar}py} K(x, x-y) d^n y. \quad \blacksquare$$

We have studied the convolution product $Q(\psi,\phi) = \operatorname{Wig}(\psi,\phi) * \theta$ for $\psi \in \mathcal{S}(\mathbb{R}^n)$ and $\theta \in \mathcal{S}(\mathbb{R}^{2n})$ (in which case $Q\psi \in \mathcal{S}(\mathbb{R}^{2n})$). It can of course also be defined under less stringent conditions. Here is a simple result for $Q\psi = Q(\psi,\psi)$:

Proposition 12 *The quasi-distribution $Q\psi = \operatorname{Wig}\psi * \theta$ is well-defined for $\psi \in L^2(\mathbb{R}^n)$ and $\theta \in L^1(\mathbb{R}^n)$, and we have in this case $Q\psi \in L^2(\mathbb{R}^{2n})$.*

Proof The condition $\psi \in L^2(\mathbb{R}^n)$ implies that $\operatorname{Wig}\psi \in L^2(\mathbb{R}^{2n})$ (Proposition 6), hence $\operatorname{Wig}\psi * \theta \in L^2(\mathbb{R}^{2n})$ by Young's theorem. ■

We refer to Boggiatto et al. [1] for a discussion of other conditions. For a general analysis of existence theorems for convolution products, see Katznelson [9].

7.2.2 The Marginal Conditions

We are mainly interested in quasi-distributions having the correct marginal properties, reproducing those of the Wigner distribution (see Proposition 3). This motivates the following terminology:

Definition 13 An element Q of the Cohen class is said to satisfy the marginal conditions if we have, for $\psi \in L^1(\mathbb{R}^n) \cap L^2(\mathbb{R}^n)$,

$$\int Q\psi(z)d^n p = |\psi(x)|^2, \quad \int Q\psi(z)d^n x = |\widehat{\psi}(p)|^2. \tag{7.27}$$

Not every element of the Cohen class satisfies the marginal conditions: see Example 15. Let us prove a necessary and sufficient condition for Q to satisfy the marginal properties.

Proposition 14 *Let $\psi \in \mathcal{S}(\mathbb{R}^n)$ and $Q\psi = \operatorname{Wig}\psi * \theta$. Assume that the mappings $x \longmapsto \theta(x,p)$ and $p \longmapsto \theta(x,p)$ are integrable. (i) We have*

$$\int Q\psi(z)d^n p = (|\psi|^2 * \alpha)(x), \quad \int Q\psi(z)d^n x = (|\widehat{\psi}|^2 * \beta)(p) \tag{7.28}$$

where the functions α and β are defined by

$$\alpha(x) = \int \theta(x,p)d^n p, \quad \beta(p) = \int \theta(x,p)d^n x.$$

(ii) The marginal properties are satisfied if and only if the symplectic Fourier transform θ_σ satisfies of the Cohen kernel θ exists and satisfies the conditions $\widehat{\theta}(x,0) = \widehat{\theta}(0,p) = (2\pi\hbar)^{-n}$; equivalently:

$$\theta_\sigma(0,p) = \theta_\sigma(x,0) = (2\pi\hbar)^{-n}. \tag{7.29}$$

Proof (i) In view of the first marginal property (7.7) satisfied by the Wigner distribution we have

$$\int \operatorname{Wig}\psi(x-x', p-p')d^n p = |\psi(x-x')|^2$$

and hence, using Fubini's theorem,

$$\begin{aligned}
\int Q\psi(z)dp &= \int \left(\int \operatorname{Wig}\psi(z-z')\theta(z')d^{2n}z' \right) d^n p \\
&= \int \left(\int \operatorname{Wig}\psi(z-z')d^n p \right) \theta(z')d^{2n}z' \\
&= \int |\psi(x-x')|^2 \left(\int \theta(z')d^n p' \right) d^n x'
\end{aligned}$$

which yields the first formula (7.28). The second formula is proven in a similar way using the second marginal property (7.7) of the Wigner transform. (ii) It suffices to show that the conditions (7.29) imply that $\alpha(x) = \delta(x)$ and $\beta(p) = \delta(p)$. Let $\widehat{\theta}$ be the usual Fourier transform of the kernel θ; we have

$$\begin{aligned}
\widehat{\theta}(x,0) &= \left(\tfrac{1}{2\pi\hbar}\right)^n \int e^{-\frac{i}{\hbar}xx'}\theta(x',p')d^n p' d^n x' \\
&= \left(\tfrac{1}{2\pi\hbar}\right)^n \int e^{-\frac{i}{\hbar}xx'} \left(\int \theta(x',p')d^n p' \right) d^n x' \\
&= \left(\tfrac{1}{2\pi\hbar}\right)^n \int e^{-\frac{i}{\hbar}xx'}\alpha(x')d^n x'
\end{aligned}$$

hence the condition $\widehat{\theta}(x, 0) = (2\pi\hbar)^{-n}$ is equivalent to $\alpha(x) = \delta(x)$. Similarly, we have $\widehat{\theta}(0, p) = (2\pi\hbar)^{-n}$; since $\theta_\sigma(z) = \widehat{\theta}(Jz)$ the conditions $\widehat{\theta}(x, 0) = \widehat{\theta}(0, p) = (2\pi\hbar)^{-n}$ are equivalent to (7.29). ■

Example 15 The Husimi distribution considered in Example 10 does not satisfy the marginal conditions, because the functions $\widehat{\theta}(x, 0)$ and $\widehat{\theta}(0, p)$ are not constants (they are themselves Gaussians, as is easily verified).

7.2.3 Generalization of Moyal's Identity

An important property of the cross-Wigner transform is the Moyal identity (7.16). A similar identity holds for Q provided that the Fourier transform of the Cohen kernel satisfies a simple condition:

Proposition 16 *Let Q be an element of the Cohen class. (i) We have*

$$\langle\langle Q(\psi, \phi)|Q(\psi', \phi')\rangle\rangle = \langle\langle \mathrm{Wig}(\psi, \phi)|\mathrm{Wig}(\psi', \phi')\rangle\rangle \tag{7.30}$$

if and only the Fourier transform $F\theta$ is such that $|F\theta(z)| = (2\pi\hbar)^{-n}$, i.e.,

$$|\theta_\sigma(z)| = (2\pi\hbar)^{-n}. \tag{7.31}$$

(ii) In particular, Moyal's identity

$$\langle\langle Q(\psi, \phi)|Q(\psi', \phi')\rangle\rangle = \left(\tfrac{1}{2\pi\hbar}\right)^n \langle\psi|\psi'\rangle\langle\phi|\phi'\rangle^* \tag{7.32}$$

holds in this case.

Proof Formulas (7.30) and (7.32) are equivalent in view of Moyal's identity (7.16). The equality $|F\theta(z)| = |F_\sigma\theta(z)|$ is obvious since $F_\sigma\theta(z) = F\theta(Jz)$. Let us prove that the identity (7.30) holds if and only if $|F\theta(z)| = (2\pi\hbar)^{2n}$. Writing $Q(\psi, \phi) = \mathrm{Wig}(\psi, \phi) * \theta$ and using Plancherel's formula together with the formula

$$FQ(\psi, \phi) = (2\pi\hbar)^n F\,\mathrm{Wig}(\psi, \phi)F\theta$$

we have, for all pairs of functions (ψ, ϕ) and (ψ', ϕ') in $\mathcal{S}(\mathbb{R}^n)$,

$$\begin{aligned}&\langle\langle Q(\psi, \phi)|Q(\psi', \phi')\rangle\rangle \\ &= (2\pi\hbar)^{2n}\langle\langle F\,\mathrm{Wig}(\psi, \phi)F\theta(z)|F\,\mathrm{Wig}(\psi', \phi')F\theta(z)\rangle\rangle\end{aligned}$$

and hence

$$\langle\langle Q(\psi,\phi)|Q(\psi',\phi')\rangle\rangle = (2\pi\hbar)^{2n}\langle\langle F\,\mathrm{Wig}(\psi,\phi)|F\,\mathrm{Wig}(\psi',\phi')|F\theta(z)|^2\rangle\rangle.$$

The equality (7.30) is thus equivalent to $|F\theta(z)|^2 = (2\pi\hbar)^{-2n}$ which proves our claim. ∎

In particular, choosing $\theta = \delta$, we recover the usual Moyal identity (7.16).

Notice that condition (7.31) does not imply the conditions (7.29), hence any Q satisfying Moyal's identity does not automatically satisfy the marginal conditions (7.28).

We will see in next chapter a fundamental counterexample: the element of the Cohen class determined by Born–Jordan quantization does not satisfy the Moyal identity (but it still satisfies the marginal conditions).

7.2.4 A Representation Result

Here is an interesting result which show that the elements of the Cohen class can be used construct pseudo-differential operators provided we have sufficient smoothness. We will see in Chap. 10 that this result applies to the Born–Jordan case.

Proposition 17 *Let $a \in \mathcal{S}'(\mathbb{R}^{2n})$ and Q be an element of the Cohen class. (i) If $Q : \mathcal{S}(\mathbb{R}^n) \times \mathcal{S}(\mathbb{R}^n) \longrightarrow \mathcal{S}(\mathbb{R}^{2n})$ there exists a unique operator $\widehat{A}_Q : \mathcal{S}(\mathbb{R}^n) \longrightarrow \mathcal{S}'(\mathbb{R}^n)$ such that we have*

$$\langle \widehat{A}_Q\psi, \phi^*\rangle = \langle\langle a, Q(\psi,\phi)\rangle\rangle \tag{7.33}$$

for all $(\psi,\phi) \in \mathcal{S}(\mathbb{R}^n) \times \mathcal{S}(\mathbb{R}^n)$. (ii) Let θ be the Cohen kernel of Q and set $\theta^\vee(z) = \theta(-z)$. The operator $\widehat{A}_Q$ is explicitly given by

$$\widehat{A}_Q = \left(\tfrac{1}{2\pi\hbar}\right)^n \int a_\sigma(z)\theta^\vee_\sigma(z)\widehat{T}(z)d^{2n}z \tag{7.34}$$

where $\theta^\vee_\sigma$ is the symplectic Fourier transform of $\theta^\vee$. (iii) Equivalently,

$$\widehat{A}_Q = \left(\tfrac{1}{\pi\hbar}\right)^n \int (a*\theta^\vee)(z)\widehat{T}_{\mathrm{GR}}(z)d^{2n}z. \tag{7.35}$$

Proof Observe that if $\phi,\psi \in \mathcal{S}(\mathbb{R}^n)$ then $Q(\psi,\phi) \in \mathcal{S}(\mathbb{R}^{2n})$, so that the right-hand side in (7.33) is well defined when $a \in \mathcal{S}'(\mathbb{R}^{2n})$. We assume that $a \in \mathcal{S}(\mathbb{R}^{2n})$ (the general case follows by duality). Since by definition

$$\langle \widehat{A}_Q\psi, \phi^*\rangle = \int a(z)(W(\psi,\phi) * \theta(z))d^{2n}z$$

we must show that the operator

$$\widehat{A} = \left(\tfrac{1}{2\pi\hbar}\right)^n \int a_\sigma(z)\theta_\sigma^\vee(z)\widehat{T}(z)d^{2n}z \tag{7.36}$$

is identical to $\widehat{A}_Q$, that is

$$\langle \widehat{A}_Q\psi, \phi^*\rangle = \int a(z)(W(\psi,\phi) * \theta)(z)d^{2n}z$$

for all $\phi, \psi \in \mathcal{S}(\mathbb{R}^n)$. In view of (7.36) we have

$$\begin{aligned}\langle \widehat{A}_Q\psi, \phi^*\rangle &= \left(\tfrac{1}{2\pi\hbar}\right)^n \int a_\sigma(z_0)\left(\int \theta_\sigma(z_0)\widehat{T}(z_0)\psi(x)\phi^*(x)d^nx\right)d^{2n}z_0 \\ &= \left(\tfrac{1}{2\pi\hbar}\right)^n \int a_\sigma(z_0)\theta_\sigma(z_0)\langle\phi|\widehat{T}(z_0)\psi\rangle d^{2n}z_0.\end{aligned}$$

In view of the definition (7.11) of the cross-ambiguity function

$$\langle\phi|\widehat{T}(z_0)\psi\rangle = \langle\widehat{T}(-z_0)\phi|\psi\rangle = (2\pi\hbar)^n\operatorname{Amb}(\psi,\phi)(-z_0)$$

and hence

$$\begin{aligned}\langle \widehat{A}_Q\psi, \phi^*\rangle &= \int a_\sigma(z_0)\theta_\sigma^\vee(z_0)\operatorname{Amb}(\psi,\phi)(-z_0)d^{2n}z_0 \\ &= \int a_\sigma(z_0)(\theta_\sigma\operatorname{Amb}(\psi,\phi))(-z_0)d^{2n}z_0\end{aligned}$$

where $\theta_\sigma^\vee(z) = \theta_\sigma(-z)$. Using the Plancherel formula for the symplectic Fourier transform, and using the relation $F_\sigma a_\sigma = a$ this equality becomes

$$\begin{aligned}\langle \widehat{A}_Q\psi, \phi^*\rangle &= \int a(z_0)F_\sigma(\theta_\sigma\operatorname{Amb}(\psi,\phi))(z_0)d^{2n}z_0 \\ &= (2\pi\hbar)^{-n}\int a(z_0)(F_\sigma\theta_\sigma * F_\sigma\operatorname{Amb}(\psi,\phi))(z_0)d^{2n}z_0;\end{aligned}$$

and using the relation $F_\sigma\theta_\sigma = \theta$ this yields

$$\begin{aligned}\langle \widehat{A}_Q\psi, \phi^*\rangle &= (2\pi\hbar)^{-n}\int a(z_0)(\theta * \operatorname{Wig}(\psi,\phi))(z_0)d^{2n}z_0 \\ &= \int a(z_0)Q(\psi,\phi)(z_0)d^{2n}z_0\end{aligned}$$

as we set out to prove. ∎

We will write

$$\widehat{T}_Q(z) = \theta_\sigma^\vee(z)\widehat{T}(z); \tag{7.37}$$

with this notation the operator (7.34) becomes

$$\widehat{A}_Q = \left(\tfrac{1}{2\pi\hbar}\right)^n \int a_\sigma(z)\widehat{T}_Q(z)d^{2n}z. \tag{7.38}$$

References

1. P. Boggiatto, E. Carypis, A. Oliaro, Local uncertainty principles for the Cohen class. J. Math. Anal. Appl. **419**(2), 1004–1022 (2014)
2. L. Cohen, Can quantum mechanics be formulated as a classical probability theory? Philos. Sci. **33**(4), 317–322 (1966)
3. G.B. Folland, *Harmonic Analysis in Phase Space, Annals of Mathematics Studies* (Princeton University Press, Princeton, 1989)
4. M. de Gosson, *Symplectic Geometry and Quantum Mechanics*. Operator Theory: Advances and Applications. Advances in Partial Differential Equations, vol. 166 (Birkhäuser, Basel 2006)
5. M. de Gosson, *Symplectic Methods in Harmonic Analysis and in Mathematical Physics*, vol. 7 (Springer Science & Business Media, 2011)
6. K. Gröchenig, *Foundations of Time-Frequency Analysis* (Birkhäuser, Boston, 2000)
7. R.L. Hudson, When is the Wigner quasi-probability density non-negative. Rep. Math. Phys. **6**, 249–252 (1974)
8. A.J.E.M. Janssen, A note on Hudson's theorem about functions with nonnegative Wigner distributions. SIAM J. Math. Anal. **15**(1), 170–176 (1984)
9. Y. Katznelson, *An Introduction to Harmonic Analysis* (Dover, 1976)
10. W.H. Zurek, Sub-Planck structure in phase space and its relevance for quantum decoherence. Nature **412**(6848), 712–717 (2001)

Chapter 8
Born–Jordan Quantization

This chapter is in a sense the keystone of this book. Using the theory of the Cohen class previously studied we give a first working definition of Born–Jordan quantization by selecting a particular Cohen kernel. We state and prove some important properties of the associated Born–Jordan operators, and discuss some unexpected properties of these operators; for instance we will show that Born–Jordan quantization is not one-to-one: the zero operator is the quantization of infinitely many classical phase space functions. Another approach, based on Shubin's theory of pseudo-differential operators, will be developed in the forthcoming chapters.

8.1 The Born–Jordan Kernel θ_{BJ}

Recall that the elements of the Cohen class are sesquilinear forms $Q : \mathcal{S}(\mathbb{R}^n) \times \mathcal{S}(\mathbb{R}^n) \longrightarrow \mathcal{S}'(\mathbb{R}^{2n})$ related to the Wigner cross-transform by the formula

$$Q(\psi, \phi) = \mathrm{Wig}(\psi, \phi) * \theta \tag{8.1}$$

where θ (the "Cohen kernel") is a tempered distribution on $\mathbb{R}^{2n}$.

8.1.1 Definition and First Properties

Born–Jordan quantization corresponds to a very particular and interesting element of the Cohen class, associated with what we call the "Born–Jordan kernel":

Definition 1 The Born–Jordan kernel is the symplectic Fourier transform

$$\theta_{\mathrm{BJ}} = \left(\tfrac{1}{2\pi\hbar}\right)^n F_\sigma \chi_{\mathrm{BJ}} \tag{8.2}$$

M.A. de Gosson, *Born–Jordan Quantization*, Fundamental Theories of Physics 182, DOI 10.1007/978-3-319-27902-2_8

of the function χ_{BJ} defined by

$$\chi_{\mathrm{BJ}}(x,p)=\operatorname{sinc}\left(\frac{px}{2\hbar}\right) \tag{8.3}$$

where sinc is the "cardinal sine" function.

The cardinal sine function[1] is defined by

$$\operatorname{sinc}x=\frac{\sin x}{x} \text{ for } x\neq 0\,,\ \operatorname{sinc}(0)=1.$$

It follows that

$$\partial_z^\alpha \chi_{\mathrm{BJ}}\in C^\infty(\mathbb{R}^{2n})\cap L^\infty(\mathbb{R}^{2n})$$

for all multi-indices $\alpha\in\mathbb{N}^{2n}$.

The Born–Jordan kernel θ_{BJ} is a real function (this easily follows using the definition of the symplectic Fourier transform and the fact that χ_{BJ} is an even real function).

Notice that we could have chosen as well the usual Fourier transform to define the Born–Jordan kernel:

$$\theta_{\mathrm{BJ}}=\left(\tfrac{1}{2\pi\hbar}\right)^n F\chi_{\mathrm{BJ}};$$

in fact, since $\sigma(z,z')=Jz\cdot z'$ we have

$$\begin{aligned}F_\sigma\chi_{\mathrm{BJ}}(z)&=\left(\tfrac{1}{2\pi\hbar}\right)^n\int e^{-\frac{i}{\hbar}Jz\cdot z'}\chi_{\mathrm{BJ}}(z')d^{2n}z'\\&=\left(\tfrac{1}{2\pi\hbar}\right)^n\int e^{-\frac{i}{\hbar}zz'}\chi_{\mathrm{BJ}}(Jz)d^{2n}z\end{aligned}$$

and hence $F_\sigma\chi_{\mathrm{BJ}}=F\chi_{\mathrm{BJ}}$ since $\chi_{\mathrm{BJ}}(Jz)=\chi_{\mathrm{BJ}}(z)$.

The cardinal sine function is one of the well-studied special functions from classical real analysis; we list below a few of its properties. The definition above, together with our previous discussion of the Cohen class motivates the introduction of the following objects:

Definition 2 (i) We denote by $\mathrm{Wig}_{\mathrm{BJ}}$ the element of the Cohen class corresponding to the choice $\theta=\theta_{\mathrm{BJ}}$:

$$\mathrm{Wig}_{\mathrm{BJ}}(\psi,\phi)=\mathrm{Wig}(\psi,\phi)*\theta_{\mathrm{BJ}}; \tag{8.4}$$

[1] It is sometimes called the "unnormalized sinc function", its "normalized" version being then $\sin(\pi x)/\pi x$ for $x\neq 0$; see formula (8.12).

we call $\mathrm{Wig}_{\mathrm{BJ}}\psi = \mathrm{Wig}_{\mathrm{BJ}}(\psi, \psi)$ the Born–Jordan quasi-distribution, or transform.
(ii) The function

$$\mathrm{Amb}_{\mathrm{BJ}}(\psi, \phi) = \mathrm{Amb}(\psi, \phi)\chi_{\mathrm{BJ}} \tag{8.5}$$

is called the Born–Jordan cross-ambiguity transform.

Notice that by the properties of the symplectic Fourier transform and formulas (7.14) in Proposition 5 in Chap. 7, and (8.2) we have

$$\mathrm{Amb}_{\mathrm{BJ}}(\psi, \phi) = F_\sigma \mathrm{Wig}_{\mathrm{BJ}}(\psi, \phi). \tag{8.6}$$

In fact, by definition (8.2) of χ_{BJ},

$$\begin{aligned} F_\sigma \mathrm{Wig}_{\mathrm{BJ}}(\psi, \phi) &= (2\pi\hbar)^n F_\sigma \mathrm{Wig}_{\mathrm{BJ}}(\psi, \phi) F_\sigma \theta_{\mathrm{BJ}} \\ &= \mathrm{Amb}(\psi, \phi)(x, p)\chi_{\mathrm{BJ}}(x, p). \end{aligned}$$

Let us now prove that $\mathrm{Wig}_{\mathrm{BJ}}$ satisfies the marginal conditions:

Proposition 3 *The function* $\mathrm{Wig}_{\mathrm{BJ}}\psi = \mathrm{Wig}\psi * \theta_{\mathrm{BJ}}$ *satisfies, for* $\psi \in L^1(\mathbb{R}^n) \cap L^2(\mathbb{R}^n)$*, the marginal conditions*

$$\int \mathrm{Wig}_{\mathrm{BJ}}\psi(z) d^n p = |\psi(x)|^2, \int \mathrm{Wig}_{\mathrm{BJ}}\psi(z) d^n x = |\widehat{\psi}(p)|^2. \tag{8.7}$$

Proof In view of Proposition 14 in Chap. 7 we have to show that

$$\widehat{\theta}_{\mathrm{BJ}}(0, p) = \widehat{\theta}_{\mathrm{BJ}}(x, 0) = (2\pi\hbar)^{-n}.$$

Now, $\theta_{\mathrm{BJ}} = (2\pi\hbar)^{-n} F\chi_{\mathrm{BJ}}$ hence $\widehat{\theta}_{\mathrm{BJ}}(z) = (2\pi\hbar)^{-n}\chi_{\mathrm{BJ}}(-z)$, that is $\widehat{\theta}_{\mathrm{BJ}}(z) = (2\pi\hbar)^{-n} \operatorname{sinc}(px/2\hbar)$. The result follows since $\operatorname{sinc}(0) = 1$. ■

Note that, as for the usual cross-Wigner transform, we have the conjugation identity

$$\mathrm{Wig}_{\mathrm{BJ}}(\psi, \phi)^* = \mathrm{Wig}_{\mathrm{BJ}}(\phi, \psi) \tag{8.8}$$

since θ_{BJ} is a real function, and hence

$$\mathrm{Wig}_{\mathrm{BJ}}(\psi, \phi)^* = \mathrm{Wig}(\psi, \phi)^* * \theta_{\mathrm{BJ}} = \mathrm{Wig}_{\mathrm{BJ}}(\phi, \psi) * \theta_{\mathrm{BJ}}.$$

It readily follows from the conjugation formula (7.13) that

$$\mathrm{Amb}_{\mathrm{BJ}}(\psi, \phi)^* = \mathrm{Amb}_{\mathrm{BJ}}(\phi^\vee, \psi^\vee) \tag{8.9}$$

where $\psi^\vee(x) = \psi(-x)$.

8.1.2 The Moyal Identity Is Not Satisfied

Recall (Proposition 16 in Chap. 7) that an element of the Cohen class satisfies the generalized Moyal identity

$$\langle\langle Q(\psi,\phi)|Q(\psi',\phi')\rangle\rangle = \left(\tfrac{1}{2\pi\hbar}\right)^n \langle\psi|\psi'\rangle\langle\phi|\phi'\rangle^* \tag{8.10}$$

if and only it is of the type $Q(\psi,\phi) = \mathrm{Wig}(\psi,\phi) * \theta$ where the Cohen kernel satisfies $|F_\sigma\theta(z)| = (2\pi\hbar)^n$. In the Born–Jordan case we have

$$F_\sigma\theta(z) = F_\sigma\theta_{\mathrm{BJ}}(z) = \mathrm{sinc}\left(\frac{px}{2\hbar}\right)$$

hence $\mathrm{Wig}_{\mathrm{BJ}}(\phi,\psi)$ cannot satisfy the Moyal identity. One can actually see this without invoking Proposition 16 in Chap. 7 using the following indirect argument.[2] Suppose indeed that

$$\langle\langle \mathrm{Wig}_{\mathrm{BJ}}(\psi,\phi)|\mathrm{Wig}_{\mathrm{BJ}}(\psi',\phi')\rangle\rangle = \left(\tfrac{1}{2\pi\hbar}\right)^n \langle\psi|\psi'\rangle\langle\phi|\phi'\rangle^*$$

for all ψ, ϕ, etc. Then, in particular

$$\langle\langle \mathrm{Wig}_{\mathrm{BJ}}\psi|\mathrm{Wig}_{\mathrm{BJ}}\psi\rangle\rangle = \left(\tfrac{1}{2\pi\hbar}\right)^n |\langle\psi|\psi\rangle|^2$$

Using Plancherel's formula and the definition (8.6) of the Born–Jordan cross-ambiguity function, this equality can be rewritten

$$\langle\langle \mathrm{Amb}_{\mathrm{BJ}}\psi|\mathrm{Amb}_{\mathrm{BJ}}\psi\rangle\rangle = \left(\tfrac{1}{2\pi\hbar}\right)^n |\langle\psi|\psi\rangle|^2$$

and using the formula (8.5) this is in turn equivalent to

$$\langle\langle (\mathrm{Amb}\psi)\chi_{\mathrm{BJ}}|(\mathrm{Amb}\psi)\chi_{\mathrm{BJ}}\rangle\rangle = \left(\tfrac{1}{2\pi\hbar}\right)^n |\langle\psi|\psi\rangle|^2.$$

The Moyal identity (7.17) for the usual cross-ambiguity transform then implies that we have

$$\langle\langle (\mathrm{Amb}\psi)\chi_{\mathrm{BJ}}|(\mathrm{Amb}\psi)\chi_{\mathrm{BJ}}\rangle\rangle = \langle\langle \mathrm{Amb}\psi|\mathrm{Amb}\psi\rangle\rangle$$

or, equivalently,

$$\int |\mathrm{Amb}\psi(z)|^2(1-\chi_{\mathrm{BJ}}(z)^2)d^{2n}z \leq 0.$$

[2] I thank Fabio Nicola for having suggested this argument.

Now,

$$\chi_{\mathrm{BJ}}(z)^2 = \operatorname{sinc}(px/2\hbar)^2 \leq 1$$

hence we must have $\operatorname{Amb}\psi(z) = 0$ for all ψ, which is not true. This contradiction shows that $\mathrm{Wig}_{\mathrm{BJ}}(\phi, \psi)$ cannot satisfy the Moyal identity.

8.1.3 *Some Properties of the* sinc *Function*

Here are a few elementary properties of sinc. First, using the Taylor expansion of $\sin x$ at the origin one immediately gets

$$\operatorname{sinc} x = \sum_{k=0}^{\infty} (-1)^k \frac{x^{2k}}{(2k+1)!}, \tag{8.11}$$

the series being convergent for all values of the real variable x; its extension to the complex domain defines an entire function of the variable $\zeta = x + iy$. The integrals of sinc and sinc^2 over the real line are equal:

$$\int_{-\infty}^{\infty} \operatorname{sinc} x dx = \int_{-\infty}^{\infty} \operatorname{sinc}^2 x dx = \pi. \tag{8.12}$$

The sinc function can be written in many ways as an infinite product; for instance

$$\operatorname{sinc}(x) = \prod_{k=1}^{\infty} \left(1 - \frac{x^2}{k^2\pi^2}\right). \tag{8.13}$$

It also satisfies the convolution identity

$$\int_{-\infty}^{\infty} \operatorname{sinc}(\pi(x-y)) \operatorname{sinc}(\pi y) dy = \operatorname{sinc}(\pi x). \tag{8.14}$$

The reciprocal $(\operatorname{sinc} x)^{-1} = x/\sin x$ is defined for all $x \neq N\pi$ (N an integer); it is an analytic function with a convergent series expansion

$$(\operatorname{sinc} x)^{-1} = \sum_{k=0}^{\infty} \frac{(-1)^{k-1} 2(2^{2k-1}-1)}{(2k)!} B_{2k} x^{2k} \tag{8.15}$$

for $|x| < \pi$; the constants B_{2k} are the Bernoulli numbers of order $2k$; recall that the Bernoulli numbers B_n are the real numbers defined by the power series

$$\frac{x}{e^x - 1} = \sum_{n=0}^{\infty} B_n \frac{x^n}{n!}. \tag{8.16}$$

8.2 Born–Jordan Operators

The results above allow us to give a first rather elegant definition of Born–Jordan quantization.

8.2.1 Definition and First Properties

Recall that a Weyl operator $\widehat{A} = \mathrm{Op}_{\mathrm{W}}(a)$ is linked to the cross-Wigner cross-ambiguity transforms by the formulae

$$\int \widehat{A}\psi(x)\phi^*(x)d^n x = \int a(z)\mathrm{Wig}(\psi, \phi)(z)d^{2n} z \tag{8.17}$$

and

$$\int \widehat{A}\psi(x)\phi^*(x)d^n x = \int a_\sigma(z)\mathrm{Amb}(\psi, \phi)(-z)d^{2n} z. \tag{8.18}$$

Since these formulas uniquely define the operator $\widehat{A} = \mathrm{Op}_{\mathrm{W}}(a)$, this suggests that we define Born–Jordan operators by replacing $\mathrm{Wig}(\psi, \phi)$ and $\mathrm{Amb}(\psi, \phi)$ with $\mathrm{Wig}_{\mathrm{BJ}}(\psi, \phi)$ and $\mathrm{Amb}_{\mathrm{BJ}}(\psi, \phi)$.

Definition 4 Let $a \in \mathcal{S}'(\mathbb{R}^{2n})$ be a symbol. The Born–Jordan operator $\widehat{A}_{\mathrm{BJ}} = \mathrm{Op}_{\mathrm{BJ}}(a)$ is the unique operator $\mathcal{S}(\mathbb{R}^n) \longrightarrow \mathcal{S}'(\mathbb{R}^n)$ such that for $(\psi, \phi) \in \mathcal{S}(\mathbb{R}^n) \times \mathcal{S}(\mathbb{R}^n)$

$$\langle \widehat{A}_{\mathrm{BJ}}\psi, \phi^* \rangle = \langle\langle a, \mathrm{Wig}_{\mathrm{BJ}}(\psi, \phi)\rangle\rangle;$$

in integral notation

$$\int \widehat{A}_{\mathrm{BJ}}\psi(x)\phi^*(x)d^n x = \int a(z)\mathrm{Wig}_{\mathrm{BJ}}(\psi, \phi)(z)d^{2n} z. \tag{8.19}$$

Equivalently,

$$\int \widehat{A}_{\mathrm{BJ}}\psi(x)\phi^*(x)d^nx = \int a_\sigma(z)\mathrm{Amb}_{\mathrm{BJ}}(\psi,\phi)(-z)d^{2n}z. \tag{8.20}$$

The equivalence of definitions (8.19) and (8.20) follows from Plancherel's formula (6.23) and formulas (8.8)–(8.9):

$$\begin{aligned}\int a(z)\mathrm{Wig}_{\mathrm{BJ}}(\psi,\phi)(z)d^{2n}z &= \int a_\sigma(z)F_\sigma\mathrm{Wig}_{\mathrm{BJ}}(\phi,\psi)(-z)d^{2n}z\\ &= \int a_\sigma(z)\mathrm{Amb}_{\mathrm{BJ}}(\psi,\phi)(-z)d^{2n}z.\end{aligned}$$

While to a given symbol there corresponds a unique Born–Jordan operator, the converse is not true: the correspondence $a \longrightarrow \mathrm{Op}_{\mathrm{BJ}}(a)$ is not one-to-one; we will discuss this important property in Sect. 8.3.1.

The following explicit result is the analogue in the Born–Jordan case of the continuous harmonic representation (6.40) of a Weyl operator in terms of the Heisenberg operators or Weyl's characteristic operator (defined by formula (6.4)):

Proposition 5 *Let $a \in \mathcal{S}'(\mathbb{R}^{2n})$ and $\psi \in \mathcal{S}(\mathbb{R}^n)$. We have*

$$\widehat{A}_{\mathrm{BJ}}\psi(x) = \left(\tfrac{1}{2\pi\hbar}\right)^n \int a_\sigma(z_0)\widehat{T}_{\mathrm{BJ}}(z_0)\psi(x)d^{2n}z_0 \tag{8.21}$$

where $\widehat{T}_{\mathrm{BJ}}(z_0) = \widehat{T}(z_0)\chi_{\mathrm{BJ}}(z_0)$, that is

$$\widehat{T}_{\mathrm{BJ}}(z_0) = \widehat{T}(z_0)\,\mathrm{sinc}\left(\frac{p_0x_0}{2\hbar}\right); \tag{8.22}$$

equivalently

$$\widehat{A}_{\mathrm{BJ}}\psi(x) = \left(\tfrac{1}{2\pi\hbar}\right)^n \int Fa(z_0)\widehat{M}_{\mathrm{BJ}}(z_0)(z_0)\psi(x)d^{2n}z_0 \tag{8.23}$$

where

$$\widehat{M}_{\mathrm{BJ}}(z_0) = \widehat{M}(z_0)\,\mathrm{sinc}\left(\frac{p_0x_0}{2\hbar}\right) \tag{8.24}$$

and $\widehat{M}(z_0) = e^{\frac{i}{\hbar}(x_0\widehat{x}+p_0\widehat{p})}$ is Weyl's characteristic operator.

Proof We assume that $a \in \mathcal{S}(\mathbb{R}^{2n})$ (the general case follows by duality) and that $(\psi,\phi) \in \mathcal{S}(\mathbb{R}^n) \times \mathcal{S}(\mathbb{R}^n)$. Since by definition

$$\langle \widehat{A}\psi, \phi^*\rangle = \int a(z)\mathrm{Wig}_{\mathrm{BJ}}(\psi,\phi)(z)d^{2n}z$$

we must show that the operator

$$\widehat{A} = \left(\tfrac{1}{2\pi\hbar}\right)^n \int a_\sigma(z)\widehat{T}_{\mathrm{BJ}}(z)d^{2n}z$$

satisfies the relation

$$\langle \widehat{A}\psi, \phi^* \rangle = \int a(z)\mathrm{Wig}_{\mathrm{BJ}}(\psi, \phi)(z)d^{2n}z \tag{8.25}$$

for all ϕ and ψ in $\mathcal{S}(\mathbb{R}^n)$. By definition of $\widehat{A}$ we have

$$\begin{aligned}
\langle \widehat{A}\psi, \phi^* \rangle &= \left(\tfrac{1}{2\pi\hbar}\right)^n \int a_\sigma(z_0) \left(\int \widehat{T}_{\mathrm{BJ}}(z_0)\psi(x)\phi^*(x)d^n x \right) d^{2n}z_0 \\
&= \left(\tfrac{1}{2\pi\hbar}\right)^n \int a_\sigma(z_0)\chi_{\mathrm{BJ}}(z_0) \left(\int \widehat{T}(z_0)\psi(x)\phi^*(x)d^n x \right) d^{2n}z_0 \\
&= \left(\tfrac{1}{2\pi\hbar}\right)^n \int a_\sigma(z_0)\chi_{\mathrm{BJ}}(z_0)\langle \phi|\widehat{T}(z_0)\psi\rangle d^{2n}z_0.
\end{aligned}$$

Taking the definition (7.11) of the cross-ambiguity function into account, we have

$$\langle \phi|\widehat{T}(z_0)\psi\rangle = \langle \widehat{T}(-z_0)\phi|\psi\rangle = (2\pi\hbar)^n \mathrm{Amb}(\psi, \phi)(-z_0)$$

and hence

$$\langle \widehat{A}\psi, \phi^* \rangle = \int a_\sigma(z_0)\chi_{\mathrm{BJ}}(z_0)\mathrm{Amb}(\psi, \phi)(-z_0)d^{2n}z_0.$$

Using the Plancherel formula (6.23) for the symplectic Fourier transform, we can rewrite this equality as

$$\langle \widehat{A}\psi, \phi^* \rangle = \int a(z_0)F_\sigma(\chi_{\mathrm{BJ}}(z_0)\mathrm{Amb}(\psi, \phi))(z_0)d^{2n}z_0$$

where we have used the fact that $F_\sigma a_\sigma = a$ since F_σ is an involution. In view of the definition (8.2),

$$\begin{aligned}
F_\sigma(\chi_{\mathrm{BJ}}\mathrm{Amb}(\psi, \phi)) &= (2\pi\hbar)^{-n} F_\sigma\chi_{\mathrm{BJ}} * F_\sigma\mathrm{Amb}(\psi, \phi) \\
&= (2\pi\hbar)^{-n} F_\sigma\chi_{\mathrm{BJ}} * \mathrm{Wig}(\psi, \phi) \\
&= \theta_{\mathrm{BJ}} * \mathrm{Wig}(\psi, \phi)
\end{aligned}$$

the last equality in view of (8.4). Summarizing, we have

$$\langle \widehat{A}\psi, \phi^* \rangle = \int a(z_0)\mathrm{Wig}_{\mathrm{BJ}}(\psi, \phi)(z_0)d^{2n}z_0$$

which we set out to prove. Formula (8.23) is equivalent to (8.21) since $\widehat{M}(z_0) = \widehat{T}(-Jz_0)$ and hence $\widehat{M}_{\mathrm{BJ}}(z_0) = \widehat{T}_{\mathrm{BJ}}(-Jz_0)$. ∎

We remark that a similar construction can actually be made for every element of the Cohen class: if $Q(\psi, \phi) = \mathrm{Wig}(\psi, \phi) * \theta$ one defines as above a generalized Heisenberg operator $\widehat{T}_\theta(z_0)$ and one then considers the operator

$$\widehat{A}_\theta = \left(\tfrac{1}{2\pi\hbar}\right)^n \int a_\sigma(z_0)\widehat{T}_\theta(z_0)d^{2n}z_0.$$

8.2.2 *The Relation Between Born–Jordan and Weyl Operators*

The following immediate consequence of Proposition 5 makes explicit the link between the Weyl operator $\mathrm{Op_W}(a)$ and the Born–Jordan operator $\mathrm{Op_{BJ}}(a)$.

Corollary 6 *Let $a \in \mathcal{S}'(\mathbb{R}^{2n})$. The Born–Jordan operator $\widehat{A}_{\mathrm{BJ}} = \mathrm{Op_{BJ}}(a)$ is the Weyl operator $\widehat{B} = \mathrm{Op_W}(b)$ with symbol $b = a * \theta_{\mathrm{BJ}} \in \mathcal{S}'(\mathbb{R}^{2n})$ where θ_{BJ} is defined by* (8.2):

$$\widehat{A}_{\mathrm{BJ}} = \mathrm{Op_W}(a * \theta_{\mathrm{BJ}}) \tag{8.26}$$

that is

$$\widehat{A}_{\mathrm{BJ}} = \left(\tfrac{1}{\pi\hbar}\right)^n \int (a * \theta_{\mathrm{BJ}})(z)\widehat{T}_{GR}(z)d^{2n}z. \tag{8.27}$$

Equivalently, the covariant Weyl symbol of $\widehat{A}_{\mathrm{BJ}}$ is the function

$$b_\sigma = a_\sigma \chi_{\mathrm{BJ}} \tag{8.28}$$

that is

$$\widehat{A}_{\mathrm{BJ}} = \left(\tfrac{1}{2\pi\hbar}\right)^n \int a_\sigma(z)\chi_{\mathrm{BJ}}(z)\widehat{T}(z)d^{2n}z.$$

Proof The formulas $b = a * \theta_{\mathrm{BJ}}$ and (8.28) are equivalent since they are symplectic Fourier transforms of each other (see the definition (8.2) of θ_{BJ}). That $a * \theta_{\mathrm{BJ}} \in \mathcal{S}'(\mathbb{R}^{2n})$ is clear. Taking into account the harmonic representation (6.37) of Weyl operators formula (8.26) implies the representation (8.27). ∎

Notice that it follows from this result that Born–Jordan operators with symbol $a \in \mathcal{S}'(\mathbb{R}^{2n})$ are *de facto* continuous operators $\mathcal{S}(\mathbb{R}^n) \longrightarrow \mathcal{S}'(\mathbb{R}^n)$ (they can be represented as Weyl operators, and the latter are continuous $\mathcal{S}(\mathbb{R}^n) \longrightarrow \mathcal{S}'(\mathbb{R}^n)$).

8.3 On the Invertibility of Born–Jordan Quantization

We are now in a position to make two fundamental remarks. The first remark is actually a question: a Born–Jordan operator, being a continuous operator $\mathcal{S}(\mathbb{R}^n) \longrightarrow \mathcal{S}'(\mathbb{R}^n)$ is *de facto* a Weyl operator (Proposition 13 in Chap. 6), and its covariant Weyl symbol is given by formula (13). How about the converse of this property? Is it true that every Weyl operator can be written as a Born–Jordan operator? The difficulty obviously comes from the formula (8.28) itself: b being given, it is not at all obvious that there exists a symbol a such that the equality (8.28) holds: we are confronted with a *division problem*. This question is fundamental, because a negative answer would imply that there are continuous operators $\mathcal{S}(\mathbb{R}^n) \longrightarrow \mathcal{S}'(\mathbb{R}^n)$ which are not Born–Jordan operators, and can thus not be "dequantized" in the Born–Jordan quantization scheme. The second remark is that it is not clear whether the Born–Jordan association $a \overset{\text{BJ}}{\longrightarrow} \widehat{A}_{\text{BJ}}$ is injective (i.e. one-to-one).

The bijectivity of Born–Jordan quantization is a serious and difficult issue. We will study a few non-trivial results below.

8.3.1 A Non-injectivity Result

We prove here a non-uniqueness result[3]; the question of invertibility, which is much more subtle, will be addressed below.

Proposition 7 *Let $a \in \mathcal{S}'(\mathbb{R}^{2n})$ and $a_{z_0}(z) = e^{-i\sigma(z,z_0)/\hbar}$. We have*

$$\operatorname{Op}_{\text{BJ}}(a) = \operatorname{Op}_{\text{BJ}}(a + a_{z_0}) \tag{8.29}$$

for all $z_0 = (x_0, p_0)$ such that $\chi_{\text{BJ}}(z_0) = 0$, that is $p_0 x_0 = 2N\pi\hbar$ ($N \in \mathbb{Z}$), $N \neq 0$.

Proof It suffices to prove that $\operatorname{Op}_{\text{BJ}}(a_{z_0}) = 0$ if $p_0 x_0 = 2N\pi\hbar$. Recall from Proposition 11 in Chap. 6 that $\operatorname{Op}_{\text{W}}(a_{z_0})$ is the Heisenberg operator $\widehat{T}(z_0)$ whose covariant symbol is $(2\pi\hbar)^n \delta(z - z_0)$. It follows from formula (8.21) that

$$\begin{aligned} \operatorname{Op}_{\text{BJ}}(a_{z_0}) &= \int \delta(z - z_0) \operatorname{sinc}\left(\tfrac{px}{2\hbar}\right) \widehat{T}(z) d^{2n}z \\ &= \int \delta(z - z_0) \operatorname{sinc}\left(\tfrac{p_0 x_0}{2\hbar}\right) d^{2n}z \\ &= 0 \end{aligned}$$

hence the result. ∎

[3] I thank V. Turunen for having pointed out this result to me.

Notice that $2N\pi\hbar = Nh$ where h is Planck's constant.

Observe that the set Σ of all z such that $\chi_{\mathrm{BJ}}(z) = 0$ consists of a family of concentric $2n-1$ dimensional sheets in phase space (when $n = 1$ they are just hyperbolas in the phase plane). The distance of the set Σ to the origin is easily calculated and one gets

$$\operatorname{dist}(\Sigma, 0) = \sqrt{4\pi\hbar}. \tag{8.30}$$

It follows from (8.29) that we have $\operatorname{Op}_{\mathrm{BJ}}(a) = 0$ for all symbols of the type

$$a(z) = \sum_{z_0 \in \Lambda} \lambda(z_0) e^{-i\sigma(z,z_0)/\hbar}$$

where Λ is any finite lattice of points $z_0 = (x_0, p_0)$ in $\mathbb{R}^{2n}$ such that $p_0 x_0 / 2\pi\hbar \in \mathbb{Z}$ and $\lambda : \Lambda \longrightarrow \mathbb{C}$. The applications of this result certainly deserves to be studied further, both in quantum mechanics and in Gabor frame theory.

8.3.2 The Case of Monomials

We assume here $n = 1$. Recall from Chap. 3 that $\mathbb{C}[x, p]$ is the polynomial ring generated by the variables x and p, and that $\mathbb{C}[\widehat{x}, \widehat{p}]$ is the Weyl algebra generated by $\widehat{x}$ and $\widehat{p}$.

We have called the operators $\widehat{A}_{\mathrm{BJ}} = \operatorname{Op}_{\mathrm{BJ}}(a)$ "Born–Jordan operators" and talked about "Born–Jordan quantization". This terminology is only justified if we prove that if $a(x, p) = x^r p^s$ where r and s are non-negative integers, then $\widehat{A}_{\mathrm{BJ}} = \operatorname{Op}_{\mathrm{BJ}}(a)$ is the operator $\mathbb{C}[x, p] \longrightarrow \mathbb{C}[\widehat{x}, \widehat{p}]$ defined by

$$\widehat{A}_{\mathrm{BJ}} = \frac{1}{s+1} \sum_{\ell=0}^{s} \widehat{p}^{s-\ell} \widehat{x}^r \widehat{p}^{\ell}$$

or by any of the equivalent formulas listed in Chap. 3. This is indeed the case, and can be done at the expense of some rather complicated calculations involving the Fourier transform of $x^r p^s$ (cf. Sect. 6.4.1 where we briefly discussed the Weyl quantization of monomials). We will therefore postpone the proof to Chap. 10, where we use an alternative pseudo-differential approach to Born–Jordan quantization making such calculations much easier. What we do here, instead, is to show that Born–Jordan operators—as defined in this chapter—are isomorphisms $\mathbb{C}[x, p] \longrightarrow \mathbb{C}[\widehat{x}, \widehat{p}]$.

Proposition 8 *The Born–Jordan quantization of polynomials is an isomorphism of vector spaces*

$$\mathrm{Op_{BJ}} : \mathbb{C}[x,p] \longrightarrow \mathbb{C}[\widehat{x},\widehat{p}].$$

Proof We are following Cordero et al. [1]. Since the Weyl transform is an isomorphism $\mathbb{C}[x,p] \longrightarrow \mathbb{C}[\widehat{x},\widehat{p}]$, every $\widehat{A} \in \mathbb{C}[\widehat{x},\widehat{p}]$ can be written $\widehat{A} = \mathrm{Op_W}(b)$ for a unique $b \in \mathbb{C}[x,p]$. This allows us to define an endomorphism T of $\mathbb{C}[\widehat{x},\widehat{p}]$ by the formula

$$T(\mathrm{Op_W}(a)) = \mathrm{Op_{BJ}}(a) = \mathrm{Op_W}(a * \chi_{\mathrm{BJ}\sigma}).$$

Let us show that T is bijective; this will prove our assertion. First, it is clear that T is injective: if $T(\mathrm{Op_W}(a)) = 0$ then $\chi_{\mathrm{BJ}}a_\sigma$ is zero as a distribution, but this is only possible if $a = 0$ since a is a polynomial, so that a_σ is supported at 0, and χ_{BJ} does not vanish in a neighborhood of 0. Let us now prove that T is surjective. Since $\mathbb{C}[x,p]$ is spanned by the monomials $b(x,p) = x^r p^s$ it is sufficient to show that there exists $a \in \mathbb{C}[x,p]$ such that $F_\sigma b = \chi_{\mathrm{BJ}} F_\sigma a$; since $F_\sigma a(z) = Fa(Jz)$ and $\chi_{\mathrm{BJ}}(Jz) = \chi_{\mathrm{BJ}}(z)$, this is equivalent to the equation $Fb(z) = \chi_{\mathrm{BJ}}(z)Fa(z)$. Since

$$Fb(z) = F(x^r \otimes p^s) = 2\pi\hbar(i\hbar)^{r+s}\delta_x^{(r)} \otimes \delta_p^{(s)}$$

the Fourier transform of a is given by

$$Fa(x,p) = 2\pi\hbar(i\hbar)^{r+s}\chi_{\mathrm{BJ}}(x,p)^{-1}\delta_x^{(r)} \otimes \delta_p^{(s)}.$$

Using the Laurent series expansion (8.15) of the sinc function we have

$$\chi_{\mathrm{BJ}}(x,p)^{-1} = \sum_{k=0}^{\infty} a_k(2\hbar)^{-2k}x^{2k}p^{2k}$$

where the coefficients are expressed in terms of the Bernoulli numbers B_n by

$$a_k = \frac{(-1)^{k-1}(2^{2k}-2)B_{2k}}{(2k)!};$$

the series is convergent in the open set $|xp| < 2\hbar\pi$. It follows that

$$\begin{aligned} Fa(x,p) &= 2\pi\hbar \sum_{k=0}^{\infty} a_k(2\hbar)^{-2k}(x^{2k}\delta_x^{(r)})(p^{2k}\delta_p^{(s)}) \\ &= 2\pi\hbar \sum_{k=0}^{n_{r,s}} a_k \frac{(2\hbar)^{-2k}r!s!}{(r-2k)!(s-2k)!}\delta_x^{(r-2k)}\delta_p^{(s-2k)} \end{aligned}$$

with $n_{r,s} = [\frac{1}{2}\min(r, s)]$ ([·] denoting the integer part). Setting

$$b_k = a_k \frac{(2\hbar)^{-2k} r!s!}{(r-2k)!(s-2k)!}$$

and noting that

$$2\pi\hbar\delta_x^{(r-2k)}\delta_p^{(s-2k)} = (i\hbar)^{-(r+s-4k)} F(x^{r-2k} p^{s-2k})$$

we have

$$a(x,p) = \sum_{k=0}^{n_{r,s}} b_k (i\hbar)^{-(r+s-4k)} x^{r-2k} p^{s-2k}$$

and hence $a \in \mathbb{C}[x, p]$. ■

8.3.3 The General Case

As we have already mentioned above, the general question of the invertibility of Born–Jordan quantization is difficult. We state here a few results which were obtained in collaboration with Elena Cordero and Fabio Nicola in [1].

It is well-known from the elementary theory of distributions that the Fourier transform of a compactly supported function (or distribution) is an entire analytic function. This statement is actually a weak version of the Paley–Wiener–Schwartz theorem (for a proof see e.g. Hörmander [2] or Vo-Khac Khoan [4]) which we state below. We denote by $B^{2n}(r)$ the closed ball in $\mathbb{R}^{2n}$ centered at the origin and with radius r, and by $\mathcal{E}'(\mathbb{R}^{2n})$ the space of compactly supported distributions on $\mathbb{R}^{2n}$; we have inclusion $\mathcal{E}'(\mathbb{R}^{2n}) \subset \mathcal{S}'(\mathbb{R}^{2n})$.

Proposition 9 *Let $a \in \mathcal{E}'(\mathbb{R}^{2n})$ have support $supp(a) \subset B^{2n}(r)$. (i) The Fourier transform Fa can be extended into an entire analytic function on $\mathbb{C}^{2n}$ and there exists constants $C > 0$, $N > 0$, such that*

$$|Fa(\zeta)| \leq C(1+|\zeta|)^N e^{\frac{r}{\hbar}|\operatorname{Im}\zeta|}; \tag{8.31}$$

where $\zeta = (\zeta_1, ..., \zeta_{2n})$ and $|\operatorname{Im}\zeta|^2 = |\operatorname{Im}\zeta_1|^2 + \cdots + |\operatorname{Im}\zeta_{2n}|^2$. (ii) Every entire analytic function a on $\mathbb{C}^{2n}$ satisfying an estimate of the type (8.31) *is the Fourier transform of some $a \in \mathcal{S}'(\mathbb{R}^{2n})$ such that $supp(a) \subset B^{2n}(r)$.*

The result above remains true if we replace the Fourier transform F with the symplectic Fourier transform F_σ.

The Paley–Wiener–Schwartz theorem suggests the following definition:

Definition 10 For $r \geq 0$ we denote by $A_r(2n)$ the subspace of $\mathcal{S}'(\mathbb{R}^{2n})$ consisting of all tempered distributions a whose symplectic Fourier transform $F_\sigma a = a_\sigma$ has support $\operatorname{supp}(a_\sigma) \subset B^{2n}(r)$. Equivalently, a satisfies an estimate

$$|a(\zeta)| \leq C(1 + |\zeta|)^N e^{\frac{r}{\hbar}|\operatorname{Im}\zeta|} \tag{8.32}$$

for some constants $C > 0$, $N > 0$.

Obviously $A_0(2) = \mathbb{C}[x, p]$, the space of polynomials in the real variables x and p. More generally, $A_0(2n)$ is the space of polynomials in the variables $x_1, ..., x_n$ and $p_1, ..., p_n$.

The following inversion result generalizes Proposition 8 about polynomials:

Proposition 11 *The linear mapping $\mathcal{S}'(\mathbb{R}^{2n}) \longrightarrow \mathcal{S}'(\mathbb{R}^{2n})$ defined by*

$$a \longmapsto \left(\tfrac{1}{2\pi\hbar}\right)^n a * \theta_{\mathrm{BJ}} \tag{8.33}$$

(or $a_\sigma \longmapsto a_\sigma \chi_{\mathrm{BJ}}$) restricts to an automorphism of $A_r(2n)$ if and only if

$$0 \leq r < \sqrt{4\pi\hbar}. \tag{8.34}$$

That is, the equation $b_\sigma = a_\sigma \chi_{\mathrm{BJ}}$ admits, for every $b \in A_r(2n)$, a unique solution $a \in A_r(2n)$ if and only if condition (8.34) *holds.*

Proof (See Cordero et al. [1].) Let us first prove the sufficiency of condition (8.34). Assume that $0 \leq r < \sqrt{4\pi\hbar}$; it follows from the equality (8.30) that the ball $B^{2n}(r)$ does not contain any zero of χ_{BJ} hence the equation $b_\sigma = a_\sigma \chi_{\mathrm{BJ}}$ admits the solution $a_\sigma = b_\sigma / \chi_{\mathrm{BJ}}$ for every $b \in A_r(2n)$, and it is clear that $a_\sigma \in \mathcal{S}'(\mathbb{R}^{2n})$. Since $\operatorname{supp}(b_\sigma) \subset B^{2n}(r)$ we also have $\operatorname{supp}(a_\sigma) \subset B^{2n}(r)$, hence $a \in A_r(2n)$. Condition (8.34) is also necessary: assume in fact that $r \geq \sqrt{4\pi\hbar}$ and choose $z_0 = (x_0, p_0)$ such that $\chi_{\mathrm{BJ}}(z_0) = 0$. Then $\widehat{T}_{\mathrm{BJ}}(z_0) = 0$ and $\operatorname{supp}(b_\sigma)$ thus contains the points z_0; hence the mapping (8.33) is not injective. ∎

In the general case we have the following surjectivity result:

Proposition 12 *The mapping* (8.33) *is a linear surjection $\mathcal{S}'(\mathbb{R}^{2n}) \longrightarrow \mathcal{S}'(\mathbb{R}^{2n})$: for every $b \in \mathcal{S}'(\mathbb{R}^{2n})$ there exists (a non-unique) $a \in \mathcal{S}'(\mathbb{R}^{2n})$ such that*

$$b = \left(\tfrac{1}{2\pi\hbar}\right)^n a * \theta_{\mathrm{BJ}}. \tag{8.35}$$

The proof of this surjectivity result is due to F. Nicola (see Theorem 6 in Cordero et al. [1]). It is highly non-trivial, and uses techniques from the theory of the division of distributions (Hörmander [3]), rewriting the equation (8.35) as $b_\sigma = a_\sigma \chi_{\mathrm{BJ}}$.

At the time of writing, it is unknown whether this result can be improved (for instance, what can one say about the solution a of (8.35) if a is a continuous function?

References

1. E. Cordero, M. de Gosson, F. Nicola, On the Invertibility of Born–Jordan quantization, preprint (2015), arXiv:1507.00144 [math.FA]
2. L. Hörmander, *The Analysis of Linear Partial Differential Operators*, vol. I (Springer, 1981)
3. L. Hörmander, On the division of distributions. Ark. Mat. **3**, 555–5568 (1958)
4. V.-K. Khoan, *Distributions, Analyse de Fourier, Opérateurs aux Dérivées Partielles, Tome 2* (Vuibert, France, 1972)

Chapter 9
Shubin's Pseudo-Differential Calculus

In this rather technical chapter we study pseudo-differential calculus from the point of view developed in Shubin [5] (also see Chap. 14 in de Gosson [3] for a review of the essentials of Shubin's theory). Shubin's operators are generalizations of Weyl operators obtained by replacing the mid-point term $\frac{1}{2}(x+y)$ in the formula

$$\widehat{A}\psi(x) = \left(\tfrac{1}{2\pi\hbar}\right)^n \int e^{\frac{i}{\hbar}p(x-y)} a(\tfrac{1}{2}(x+y), p)\psi(y)d^n p d^n y$$

with the weighted average $(1-\tau)x + \tau y$ where τ is an arbitrary real parameter. The theory of Shubin operators (which we will often call "τ-operators") is a very convenient tool easily leading to an alternative definition of Born–Jordan operators: these are obtained by averaging the Shubin operators with the same symbol over the unit interval. This allows us at the same time to recover the Born–Jordan transform of a couple of functions without a direct recourse to the properties of the Cohen class. Shubin operators do not have themselves the physical properties qualifying them as a *bona fide* quantization (for instance, real observables do not lead to self-adjoint operators in the Shubin scheme). This pseudo-differential calculus should thus be viewed as a useful technical intermediary without direct physical significance.

9.1 Definition and First Properties

In what follows τ is a real parameter.

9.1.1 The Kernel of a τ-Operator

The integral formula (6.38) for a Weyl operator, which reads

$$\widehat{A}\psi(x) = \left(\tfrac{1}{2\pi\hbar}\right)^n \int e^{\frac{i}{\hbar}p(x-y)} a(\tfrac{1}{2}(x+y), p)\psi(y)d^n p d^n y,$$

M.A. de Gosson, *Born–Jordan Quantization*, Fundamental Theories of Physics 182, DOI 10.1007/978-3-319-27902-2_9

suggests the possibility of defining more general pseudo-differential operators by

$$\widehat{A}_\tau\psi(x) = \left(\tfrac{1}{2\pi\hbar}\right)^n \int e^{\frac{i}{\hbar}p(x-y)} a((1-\tau)x + \tau y, p)\psi(y) d^n y d^n p \tag{9.1}$$

where τ is some (arbitrary but fixed) real parameter. Note that the Weyl operators correspond to the choice $\tau = \frac{1}{2}$. Here is a rigorous definition:

Definition 1 Let $a \in S'(\mathbb{R}^{2n})$. The τ-pseudo-differential operator (or simply τ-pseudo-differential operator) with symbol a is the operator $\widehat{A}_\tau = \mathrm{Op}_\tau(a)$ with distributional kernel

$$K_\tau(x, y) = \left(\tfrac{1}{2\pi\hbar}\right)^{n/2} (F_2^{-1}a)((1-\tau)x + \tau y, x - y) \tag{9.2}$$

where F_2^{-1} is the inverse Fourier transform in the p variables; in integral notation

$$K_\tau(x, y) = \left(\tfrac{1}{2\pi\hbar}\right)^n \int e^{\frac{i}{\hbar}p(x-y)} a((1-\tau)x + \tau y, p) d^n p. \tag{9.3}$$

We will also use the notation $\widehat{A}_\tau \overset{\tau}{\longleftrightarrow} a$ or $a \overset{\tau}{\longleftrightarrow} \widehat{A}_\tau$. Notice that, conversely, the symbol is expressed in terms of the kernel by the formula

$$a(x, p) = \int e^{-\frac{i}{\hbar}py} K(x + \tau y, x - (1-\tau)y, p) d^n p. \tag{9.4}$$

In practice we will use the integral notation (9.1) for operators.

Example 2 Suppose we choose $\tau = 0$. Then

$$\widehat{A}_0\psi(x) = \left(\tfrac{1}{2\pi\hbar}\right)^n \int e^{\frac{i}{\hbar}p(x-y)} a(x, p)\psi(y) d^n y d^n p$$

which we can rewrite as

$$\widehat{A}_0\psi(x) = \left(\tfrac{1}{2\pi\hbar}\right)^{n/2} \int e^{\frac{i}{\hbar}px} a(x, p)\widehat{\psi}(p) d^n p.$$

This is the so-called "normal ordering" familiar from the theory of partial differential equations. Similarly, the choice $\tau = 1$ leads to the " antinormal ordering".

We will in a moment give alternative descriptions of τ-pseudo-differential operators by imitating the harmonic representation of Weyl operators; this will allow us to recover in the next chapter the Born–Jordan machinery previously introduced by other means.

9.1.2 *The Case of Monomials*

Let us find the τ-pseudo-differential operator corresponding to the monomial symbols $x_j^r p_j^s$ considered in the Introduction. It is of course sufficient to consider the case $n = 1$; we will write $x_j^r = x^r$ and $p_j^s = p^s$.

Proposition 3 *Let r and s be two non-negative integers. We have*

$$\mathrm{Op}_\tau(x^r p^s) = \sum_{k=0}^{r} \binom{r}{k} \tau^k (1-\tau)^{r-k} \widehat{x}^k \widehat{p}^s \widehat{x}^{r-k} \tag{9.5}$$

or, equivalently,

$$\mathrm{Op}_\tau(x^r p^s) = \sum_{k=0}^{s} \binom{s}{k} (1-\tau)^k \tau^{s-k} \widehat{p}^k \widehat{x}^r \widehat{p}^{s-k} \tag{9.6}$$

where $\widehat{x}^s \psi = x^s \psi$ and $\widehat{p}^s \psi = (-i\hbar\partial_x)^s \psi$ for $\psi \in \mathcal{S}(\mathbb{R})$.

Proof Let us set $a_{r,s}(z) = x^r p^s$; we have using the binomial formula

$$a_{r,s}(\tau x + (1-\tau)y, p) = \sum_{k=0}^{r} \binom{r}{k} \tau^k (1-\tau)^{r-k} x^k y^{r-k} p^s.$$

Setting $b_{r,s,k}(z) = x^k y^{r-k} p^s$ we have (in the sense of distributions)

$$\mathrm{Op}_\tau(b_{r,s,k})\psi(x) = \frac{x^k}{2\pi\hbar} \int_{-\infty}^{\infty} \left[\int_{-\infty}^{\infty} e^{\frac{i}{\hbar}p(x-y)} p^s dp \right] y^{r-k} \psi(y) dy.$$

Using the Fourier inversion formula

$$\frac{1}{2\pi\hbar} \int_{-\infty}^{\infty} e^{\frac{i}{\hbar}p(x-y)} p^s dp = (-i\hbar)^s \delta^{(s)}(x-y) \tag{9.7}$$

we thus have

$$\mathrm{Op}_\tau(b_{r,s,k})\psi = x^k (-i\hbar)^s \partial_x^s (x^{r-k}\psi).$$

Formula (9.5) follows inserting this expression in (3.23). To prove that this formula is equivalent to (9.6) the easiest method consists in remarking that we have the conjugation formula

$$F\mathrm{Op}_\tau(a)F^{-1} = \mathrm{Op}_{1-\tau}(a \circ J^{-1})$$

where F is the Fourier transform (this equality will be proven in Chap. 12 as a particular case of "symplectic covariance"). Since we have $a_{r,s}(J^{-1}z) = (-1)^r x^r p^s$ and, using the standard properties of the Fourier transform,

$$F\mathrm{Op}_\tau(a_{r,s})F^{-1} = (-1)^r \widehat{p}^k \widehat{x}^s \widehat{p}^{r-k}$$

formula (9.6) follows. ∎

Notice that the formulas (9.5) and (9.6) were proposed in Sect. 3.2.4 of Chap. 3 (formulas (3.16) and (3.17)) as intermediaries leading to the Born–Jordan quantization by averaging over τ from 0 to 1.

We will see (Corollary 15) that Shubin operators admit a continuous harmonic analysis of the type

$$\widehat{A}_\tau = \left(\tfrac{1}{\pi\hbar}\right)^n \int a(z)\widehat{T}_{\mathrm{GR},\tau}(z)d^{2n}z$$

or

$$\widehat{A}_\tau = \left(\tfrac{1}{2\pi\hbar}\right)^n \int a_\sigma(z)\widehat{T}_\tau(z)d^{2n}z$$

where $\widehat{T}_{\mathrm{GR},\tau}(z)$ and $\widehat{T}_\tau(z)$ are generalizations of the usual Grossmann–Royer and Heisenberg operators. This can be proven directly at the expense of rather heavy calculations; to avoid these complications we will use an indirect approach, using the notion of τ-Wigner transform.

9.2 The τ-Wigner and Ambiguity Transforms

We extend the notions of cross-Wigner and ambiguity transforms to the case of τ-dependent operators.

9.2.1 *The Operators $\widehat{T}_\tau(z_0)$ and $\widehat{T}_{\mathrm{GR},\tau}(z_0)$*

Recall (formula (6.10)) that the Heisenberg operator is defined for $\psi \in \mathcal{S}'(\mathbb{R}^n)$ and $z_0 = (x_0, p_0)$ by

$$\widehat{T}(z_0)\psi(x) = e^{\frac{i}{\hbar}(p_0x - \frac{1}{2}p_0x_0)}\psi(x - x_0). \tag{9.8}$$

Definition 4 The Heisenberg τ-operator $\widehat{T}_\tau(z_0)$ is defined by the formula

$$\widehat{T}_\tau(z_0)\psi(x) = e^{\frac{i}{\hbar}(p_0x-(1-\tau)p_0x_0)}\psi(x - x_0) \tag{9.9}$$

that is, equivalently,

$$\widehat{T}_\tau(z_0) = e^{\frac{i}{2\hbar}(2\tau-1)p_0x_0}\widehat{T}(z_0). \tag{9.10}$$

We have $\widehat{T}_{1/2}(z_0) = \widehat{T}(z_0)$, and

$$\widehat{T}_\tau(z_0)^{-1} = \widehat{T}_{1-\tau}(-z_0). \tag{9.11}$$

It is immediate to check the relations:

$$\widehat{T}_\tau(z_0)\widehat{T}_\tau(z_1) = e^{\frac{i}{\hbar}\sigma_\tau(z_0,z_1)}\widehat{T}_\tau(z_1)\widehat{T}_\tau(z_0) \tag{9.12}$$

$$\widehat{T}_\tau(z_0+z_1) = e^{-\frac{i}{2\hbar}\sigma_\tau(z_0,z_1)}\widehat{T}_\tau(z_0)\widehat{T}_\tau(z_1) \tag{9.13}$$

(they follow from the relations (6.14) and (6.15) satisfied by the Heisenberg operators), where σ_τ is the bilinear form defined by

$$\sigma_\tau(z_0, z_1) = 2(1-\tau)p_0x_1 - 2\tau p_1x_0. \tag{9.14}$$

Note that σ_τ fails to be antisymmetric if $\tau \neq \frac{1}{2}$, so it is not in general a symplectic form.

Recall (formula (6.11)) that the Grossmann–Royer operators are defined by

$$\widehat{T}_{\mathrm{GR}}(z_0)\psi(x) = \widehat{T}_{\mathrm{GR}}(z_0)\psi(x) = e^{\frac{2i}{\hbar}p_0(x-x_0)}\psi(2x_0-x).$$

Definition 5 The τ-Grossmann–Royer operator is defined by

$$\widehat{T}_{\mathrm{GR},\tau}(z)\psi = 2^{-n}F_\sigma(\widehat{T}_\tau(\cdot)\psi)^\vee \tag{9.15}$$

where $^\vee$ is the reflection operator $z \longmapsto -z$.

Example 6 When $\tau = \frac{1}{2}$ the operator $\widehat{T}_{\mathrm{GR},\tau}(z)$ is the usual Grossmann–Royer operator: in view formula (6.27) we have

$$\begin{aligned}
\widehat{T}_{\mathrm{GR},1/2}(z_0)\psi(x) &= 2^{-n}F_\sigma[\widehat{T}_{1/2}(\cdot)\psi(x)](-z_0) \\
&= 2^{-n}F_\sigma[\widehat{T}(\cdot)\psi(x)](-z_0) \\
&= \widehat{T}_{\mathrm{GR}}(z_0)\psi(x)
\end{aligned}$$

The operators $\widehat{T}_{\mathrm{GR},\tau}(z)$ and $\widehat{T}_{\mathrm{GR}}(z)$ are related by a convolution formula:

Proposition 7 *We have, for* $\tau \neq \frac{1}{2}$,

$$\widehat{T}_{\mathrm{GR},\tau}(z)\psi = \theta_{(\tau)} * \widehat{T}_{\mathrm{GR}}(z)\psi \tag{9.16}$$

where the function $\theta_{(\tau)} : \mathbb{R}^{2n} \longrightarrow \mathbb{C}$ *is defined by*

$$\theta_{(\tau)}(z) = \frac{2^n}{|2\tau - 1|^n} e^{\frac{2i}{\hbar(2\tau-1)} px}. \tag{9.17}$$

(The convolution product in (9.16) is calculated with respect to the z variable.)

Proof We have, by definition of F_σ and (9.10),

$$F_\sigma[\widehat{T}_\tau(\cdot)\psi](z) = \left(\tfrac{1}{2\pi\hbar}\right)^n \int e^{-\frac{i}{\hbar}\sigma(z,z')} e^{\frac{i}{2\hbar}(2\tau-1)p'x'} \widehat{T}(z')\psi d^{2n}z'$$

that is, using the convolution formula (6.24),

$$F_\sigma[\widehat{T}_\tau(\cdot)\psi] = (2\pi\hbar)^{-n}(F_\sigma a) * F_\sigma(\widehat{T}(\cdot)\psi)$$

where $a(z') = e^{i(2\tau-1)p'x'/2\hbar}$. Taking into account the general relation $(f * g)^\vee = f^\vee * g^\vee$ where $^\vee$ is the phase space reflection operator $z \longmapsto -z$, this equality can be rewritten

$$F_\sigma[\widehat{T}_\tau(\cdot)\psi]^\vee = (2\pi\hbar)^{-n}(F_\sigma a)^\vee * F_\sigma(\widehat{T}(\cdot)\psi)^\vee;$$

now $(F_\sigma a)^\vee = F_\sigma a$ since $a(-z') = a(z')$, hence

$$F_\sigma[\widehat{T}_\tau(\cdot)\psi]^\vee = (2\pi\hbar)^{-n}(F_\sigma a) * F_\sigma(\widehat{T}(\cdot)\psi)^\vee.$$

In view of the identity

$$F_\sigma(\widehat{T}(\cdot)\psi)^\vee(z_0) = 2^n \widehat{T}_{\mathrm{GR},\tau}(z_0)\psi$$

(formula (9.15)) it is thus sufficient to prove that

$$F_\sigma a(z) = \frac{2^n}{|2\tau - 1|^n} e^{\frac{2i}{\hbar(2\tau-1)} px} = \theta_{(\tau)}(z). \tag{9.18}$$

We have (the integrals being interpreted as distributional brackets)

$$\begin{aligned}
F_\sigma a(z) &= \left(\tfrac{1}{2\pi\hbar}\right)^n \int e^{-\frac{i}{\hbar}\sigma(z,z')} e^{\frac{i}{2\hbar}(2\tau-1)p'x'} d^n p' d^n x' \\
&= \left(\tfrac{1}{2\pi\hbar}\right)^n \int e^{-\frac{i}{\hbar}[px' - p'x - \frac{1}{2}(2\tau-1)p'x']} d^n p' d^n x' \\
&= \left(\tfrac{1}{2\pi\hbar}\right)^n \int e^{\frac{i}{\hbar}p'x} \left[\int e^{-\frac{i}{\hbar}[(p - \frac{1}{2}(2\tau-1)p')x']} d^n x'\right] d^n p' \\
&= \int e^{\frac{i}{\hbar}p'x} \delta(p - \tfrac{1}{2}(2\tau - 1)p') d^n p'.
\end{aligned}$$

Setting $p'' = p - \frac{1}{2}(2\tau - 1)p'$ in the last integral we get (9.18). ∎

9.2.2 The Associated Transforms

Boggiatto and his collaborators [1, 2] have introduced a τ-dependent Wigner transform $\mathrm{Wig}_\tau(\psi, \phi)$ (also see Toft [6]) defined by

$$\mathrm{Wig}_\tau(\psi, \phi)(z) = \left(\tfrac{1}{2\pi\hbar}\right)^n \int e^{-\frac{i}{\hbar}py}\psi(x + \tau y)\phi^*(x - (1 - \tau)y)d^n y. \tag{9.19}$$

We are going to see that $\mathrm{Wig}_\tau(\psi, \phi)$ can be obtained using the τ-Grossmann–Royer operator.

9.2.3 Definition of Wig_τ and Amb_τ

The (cross-)Wigner transform was defined in the previous chapter (formula (7.3)) in terms of the Grossmann–Royer operator. We define the τ-Wigner transform $\mathrm{Wig}_\tau(\psi, \phi)$ using the τ-Grossmann–Royer operator:

Definition 8 Let $(\psi, \phi) \in \mathcal{S}(\mathbb{R}^n) \times \mathcal{S}(\mathbb{R}^n)$. The function $(\psi, \phi) \longmapsto \mathrm{Wig}_\tau(\psi, \phi)$ defined by

$$\mathrm{Wig}_\tau(\psi, \phi)(z) = \left(\tfrac{1}{\pi\hbar}\right)^n \langle \widehat{T}_{\mathrm{GR},\tau}(z)\phi|\psi\rangle \tag{9.20}$$

is called the τ-cross-Wigner function (or transform) of (ψ, ϕ).

Using the explicit expression

$$\widehat{T}_{\mathrm{GR},\tau}(z)\psi = \theta_{(\tau)} * \widehat{T}_{\mathrm{GR}}(z)\psi \tag{9.21}$$

(formula (9.16)) valid for $\tau \neq \frac{1}{2}$, where

$$\theta_{(\tau)}(z) = \frac{2^n}{|2\tau - 1|^n} e^{\frac{2i}{\hbar(2\tau-1)}px} \tag{9.22}$$

we get the following relation between $\mathrm{Wig}_\tau(\psi, \phi)$ and $\mathrm{Wig}(\psi, \phi)$:

$$\mathrm{Wig}_\tau(\psi, \phi) = \mathrm{Wig}(\psi, \phi) * \theta_{(\tau)}. \tag{9.23}$$

As is the case for Wig, the mapping Wig_τ is a sesquilinear and continuous mapping $\mathcal{S}(\mathbb{R}^n) \times \mathcal{S}(\mathbb{R}^n) \longrightarrow \mathcal{S}(\mathbb{R}^{2n})$, in fact an element of the Cohen class (see Definition 9 in Chap. 7). When $\psi = \phi$ one writes $\mathrm{Wig}_\tau(\psi, \psi) = \mathrm{Wig}_\tau\psi$ (the τ-Wigner quasi distribution).

The following result gives the explicit expression of $\mathrm{Wig}_\tau(\psi, \phi)$ (*cf.* formula (9.19)):

Proposition 9 *We have, for* $(\psi, \phi) \in \mathcal{S}(\mathbb{R}^n) \times \mathcal{S}(\mathbb{R}^n)$ *and* $\tau \in \mathbb{R}$,

$$\mathrm{Wig}_\tau(\psi,\phi)(z) = \left(\tfrac{1}{2\pi\hbar}\right)^n \int e^{-\frac{i}{\hbar}py}\psi(x+\tau y)\phi^*(x-(1-\tau)y)d^n y \tag{9.24}$$

for $\tau \neq \frac{1}{2}$ *and* $\mathrm{Wig}_{1/2}(\psi,\phi) = \mathrm{Wig}(\psi,\phi)$.

Proof Let us set, for $\tau \neq \frac{1}{2}$,

$$\beta_\tau(z) = (2\pi\hbar)^n e^{\frac{2ipx}{\hbar(2\tau-1)}}.$$

We have

$$\mathrm{Wig}(\psi,\phi) * \beta_\tau(z) = \int \mathrm{Wig}(\psi,\phi)(z-z')\beta_\tau(z')d^{2n}z'$$

and, by definition of β_τ,

$$\mathrm{Wig}(\psi,\phi)(z-z')\beta_\tau(z') = \int e^{-\frac{i}{\hbar}py} e^{\frac{i}{\hbar}p'\left(y+\frac{2x'}{2\tau-1}\right)} \times \psi(x-x'+\tfrac{1}{2}y)\phi^*(x-x'-\tfrac{1}{2}y)d^n y.$$

Integrating this expression with respect to p' yields

$$\int \mathrm{Wig}(\psi,\phi)(z-z')\beta_\tau(z')d^n p' = \int e^{-\frac{i}{\hbar}py}\delta(y+\tfrac{2x'}{2\tau-1}) \times \psi(x-x'+\tfrac{1}{2}y)\phi^*(x-x'-\tfrac{1}{2}y)d^n y$$

that is

$$\int \mathrm{Wig}(\psi,\phi)(z-z')\beta_\tau(z')d^n p' = \int e^{-\frac{i}{\hbar}py}\delta(y+\tfrac{2x'}{2\tau-1}) \times \psi(x+\tau y)\phi^*(x-(1-\tau)y)d^n y$$

where we have used the identity

$$\delta(y+\tfrac{2x'}{2\tau-1})\psi(x-x'+\tfrac{1}{2}y)\phi(x-x'-\tfrac{1}{2}y)^* = \delta(y+\tfrac{2x'}{2\tau-1})\psi(x+\tau y)\phi^*(x-(1-\tau)y)d^n y.$$

Integrating with respect to x' and noting the identity

$$\int \delta(y+\tfrac{2x'}{2\tau-1})d^n x' = \left(\tfrac{|2\tau-1|}{2}\right)^n$$

we get

$$\operatorname{Wig}(\psi,\phi) * \beta_\tau(z) = \left(\tfrac{|2\tau-1|}{2}\right)^n \times \int e^{-\frac{i}{\hbar}py}\psi(x+\tau y)\phi^*(x-(1-\tau)y)d^n y$$

which is equivalent to (9.24). ∎

Note that formula (9.24) also makes sense for $(\psi,\phi) \in L^2(\mathbb{R}^n) \times L^2(\mathbb{R}^n)$ or $(\psi,\phi) \in \mathcal{S}(\mathbb{R}^n) \times \mathcal{S}'(\mathbb{R}^n)$ provided the integral is interpreted in the distributional sense.

In particular, when $\tau = \frac{1}{2}$, one recovers the usual cross-Wigner transform $\operatorname{Wig}(\psi,\phi)$. If $\tau = 0$ we get

$$\operatorname{Wig}_0(\psi,\phi)(z) = \left(\tfrac{1}{2\pi\hbar}\right)^{n/2} e^{-\frac{i}{\hbar}px}\psi(x)(F\phi)^*(p)$$

which is, when $\psi = \phi$, the Rihaczek–Kirkwood distribution well-known from time-frequency analysis (Gröchenig [4], Boggiatto et al. [1]); if $\tau = 1$ one gets the so-called dual Rihaczek–Kirkwood distribution $R^*(\phi,\psi)$.

Definition 5 and the theory of the (cross-)ambiguity transform developed in the previous chapters suggest:

Definition 10 Let $(\psi,\phi) \in \mathcal{S}(\mathbb{R}^n) \times \mathcal{S}(\mathbb{R}^n)$. The τ-cross ambiguity transform of (ψ,ϕ) is the function $\mathcal{S}(\mathbb{R}^n) \times \mathcal{S}(\mathbb{R}^n) \longrightarrow \mathcal{S}(\mathbb{R}^{2n})$ defined by

$$\operatorname{Amb}_\tau(\psi,\phi) = F_\sigma \operatorname{Wig}_\tau(\psi,\phi). \tag{9.25}$$

Equivalently,

$$\operatorname{Amb}_\tau(\psi,\phi)(z) = \left(\tfrac{1}{2\pi\hbar}\right)^n \langle \widehat{T}_\tau(z)\phi|\psi\rangle. \tag{9.26}$$

The equivalence of the definitions (9.25) and (9.26) follows from the definition (9.15) of $\widehat{T}_{\mathrm{GR},\tau}(z)$.

The τ-cross-ambiguity transform is related to the usual cross-ambiguity transform by a very simple formula:

Proposition 11 *Let $(\psi,\phi) \in \mathcal{S}(\mathbb{R}^n) \times \mathcal{S}(\mathbb{R}^n)$. We have*

$$\operatorname{Amb}_\tau(\psi,\phi)(z) = e^{-\frac{i}{\hbar}(2\tau-1)px}\operatorname{Amb}(\psi,\phi)(z); \tag{9.27}$$

that is

$$\operatorname{Amb}_\tau(\psi,\phi)(z) = \left(\tfrac{1}{2\pi\hbar}\right)^n e^{-\frac{i}{\hbar}(2\tau-1)px} \times \int e^{-\frac{i}{\hbar}py}\psi(y+\tau x)\phi^*(y-(1-\tau)x)d^n y.$$

Proof These formulas actually follow directly from definition (9.10) of the operator $\widehat{T}_\tau(z)$ using the definition (7.11) of the usual cross-ambiguity function; we however give here a direct independent proof. In view of the definition (9.9) of $\widehat{T}_\tau(z_0)$ we have

$$\widehat{T}_\tau(z_0)\phi(x) = e^{\frac{i}{\hbar}(p_0x-(1-\tau)p_0x_0)}\phi(x-x_0)$$

hence

$$\langle\widehat{T}_\tau(z_0)\phi|\psi\rangle = \int e^{-\frac{i}{\hbar}(p_0x'-(1-\tau)p_0x_0)}\phi^*(x'-x_0)\psi(x')d^nx';$$

setting $y = x' - \tau x_0$ we get

$$\langle\widehat{T}_\tau(z_0)\phi|\psi\rangle = e^{-\frac{i}{\hbar}(2\tau-1)p_0x_0}\int e^{-\frac{i}{\hbar}p_0y}\phi^*(y-(1-\tau)x_0)\psi(y+\tau x_0)d^ny$$

hence (9.27). ∎

Notice that we recover the usual cross-ambiguity function (7.11) and (7.12) for $\tau = \frac{1}{2}$. We will set

$$\mathrm{Amb}_\tau\psi = \mathrm{Amb}_\tau(\psi,\psi)$$

and call $\mathrm{Amb}_\tau\psi$ the *τ-ambiguity function*.

9.2.4 Properties of Wigτ and Amb$_\tau$

It is easily verified using (9.24) that the τ-cross-Wigner transform satisfies the conjugation relation

$$\mathrm{Wig}_\tau(\phi,\psi)^* = \mathrm{Wig}_{1-\tau}(\psi,\phi);$$

in particular

$$(\mathrm{Wig}_\tau\psi)^* = \mathrm{Wig}_{1-\tau}\psi$$

hence $\mathrm{Wig}_\tau\psi = \mathrm{Wig}_\tau(\psi,\psi)$ is generically real if and only if $\tau = \frac{1}{2}$. However $\mathrm{Wig}_\tau\psi$ satisfies for $\psi \in L^1(\mathbb{R}^n)\cap L^2(\mathbb{R}^n)$ the same marginal properties as $\mathrm{Wig}\psi$ (*cf.* (7.7)):

$$\int \mathrm{Wig}_\tau\psi(z)d^np = |\psi(x)|^2, \int \mathrm{Wig}_\tau\psi(z)d^nx = |F\psi(p)|^2 \tag{9.28}$$

(see Boggiatto et al. [1]); this can be seen as a particular case of the theory of the Cohen.

In Weyl calculus the introduction of the Wigner transform $\mathrm{Wig}\psi$ of a square integrable function has the following very simple and natural interpretation: it is, up to a constant factor, the Weyl symbol of the projection operator Π_ψ of $L^2(\mathbb{R}^n)$ on the ray $\{\lambda\psi : \lambda \in \mathbb{C}\}$. This interpretation extends to the τ-dependent case without difficulty:

Proposition 12 *Let $\psi \in L^2(\mathbb{R}^n)$, $\psi \neq 0$, and let Π_ψ be the orthogonal projection operator on the ray spanned by ψ. (i) We have*

$$\Pi_\psi = (2\pi\hbar)^n \operatorname{Op}_\tau(\operatorname{Wig}_\tau \psi).$$

(ii) The τ-symbol of the operator with kernel $K = \psi \otimes \phi^*$ is the function $(2\pi\hbar)^n \operatorname{Wig}_\tau(\psi, \phi)$.

Proof (i) Let $\phi \in L^2(\mathbb{R}^n)$; by definition $\Pi_\psi \phi = \langle \phi | \psi \rangle \psi$ that is

$$\Pi_\psi \phi(x) = \int \psi(x)\psi(y)^* \phi(y) d^n y$$

hence the kernel of Π_ψ is $K(x, y) = \psi(x)\psi(y)^*$. Using a partial Fourier inversion formula, formula (9.2) expressing the kernel of Π_ψ in terms of its τ-symbol π_ψ can be rewritten

$$\begin{aligned}\pi_\psi(x, p) &= \int e^{-\frac{i}{\hbar} py} K(x + \tau y, x - (1 - \tau)y) d^n y \\ &= \int e^{-\frac{i}{\hbar} py} \psi(x + \tau y)\psi^*(x - (1 - \tau)y) d^n y \\ &= (2\pi\hbar)^n \operatorname{Wig}_\tau \psi(x, p).\end{aligned}$$

The assertion (ii) is proven in a similar way replacing $\psi \otimes \psi^*$ with $\psi \otimes \phi^*$ in the argument above. ∎

We also have the following extension of the usual Moyal identity (7.16); it can actually be checked directly using Proposition 16 in Chap. 7: in fact, a straightforward calculation shows that the Fourier transform of the Cohen kernel $\theta_{(\tau)}$ is given, for $\tau \neq \frac{1}{2}$, by

$$\widehat{\theta}_{(\tau)}(z) = \left(\tfrac{1}{2\pi\hbar}\right)^n \tfrac{(2\tau-1)^n}{|2\tau-1|^n} e^{-\frac{i(2\tau-1)}{2\hbar} px}. \tag{9.29}$$

and hence $|\widehat{\theta}_{(\tau)}(z)| = (2\pi\hbar)^{-n}$, and it follows that $W_\tau(\psi, \phi)$ satisfies the Moyal identity. For the sake of self-containedness we give here a direct derivation:

Proposition 13 *Let $\langle\langle \cdot | \cdot \rangle\rangle$ be the scalar product on $L^2(\mathbb{R}^{2n})$ and $|||\cdot|||$ the associated norm. We have*

$$\langle\langle \operatorname{Wig}_\tau(\psi, \phi) | \operatorname{Wig}_\tau(\psi', \phi') \rangle\rangle = \left(\tfrac{1}{2\pi\hbar}\right)^n \langle \psi | \psi' \rangle \langle \phi | \phi' \rangle^* \tag{9.30}$$

and

$$\langle\langle \operatorname{Amb}_\tau(\psi, \phi) | \operatorname{Amb}_\tau(\psi', \phi') \rangle\rangle = \left(\tfrac{1}{2\pi\hbar}\right)^n \langle \psi | \psi' \rangle \langle \phi | \phi' \rangle^* \tag{9.31}$$

for all $\psi, \psi', \phi, \phi' \in L^2(\mathbb{R}^n)$. *Hence, in particular,*

$$|||\mathrm{Wig}_\tau(\psi,\phi)||| = |||\mathrm{Amb}_\tau(\psi,\phi)||| = \left(\tfrac{1}{2\pi\hbar}\right)^{n/2} ||\psi||\,||\phi||. \tag{9.32}$$

Proof Let us set

$$I = (2\pi\hbar)^{2n} \langle\langle \mathrm{Wig}_\tau(\psi,\phi)|\mathrm{Wig}_\tau(\psi',\phi')\rangle\rangle.$$

We have, using the explicit expression (9.24) of $\mathrm{Wig}_\tau(\psi,\phi)$,

$$I = \int e^{-\frac{i}{\hbar}p(y-y')}\psi^*(x+\tau y)\psi'(x+\tau y)$$
$$\times\, \phi(x-(1-\tau)y')\phi'^*(x-(1-\tau)y')d^nxd^npd^nyd^ny'.$$

The integral in p is equal to $(2\pi\hbar)^n\,\delta(y-y')$, hence

$$I = (2\pi\hbar)^n \int \psi^*(x+\tau y)\psi'(x+\tau y)$$
$$\times\, \phi(x-(1-\tau)y)\phi'^*(x-(1-\tau)y)d^nxd^ny.$$

Setting $u = x+\tau y$ and $v = x-(1-\tau)y$ we have $d^nud^nv = d^nxd^ny$ and hence

$$I = (2\pi\hbar)^n \int \psi^*(u)\psi'(u)\phi(v)\phi'^*(v)d^nud^nv$$
$$= (2\pi\hbar)^n \langle\psi|\psi'\rangle\langle\phi|\phi\rangle^*$$

which proves (9.30); formula (9.32) follows. Formula (9.31) follows from (9.30) using Plancherel's formula (6.22). ■

We will see (formula (9.45)) that the (formal) adjoint $\widehat{A}_\tau^\dagger$ of the τ-pseudo-differential operator $\widehat{A}_\tau = \mathrm{Op}_\tau(a)$ is given by the formula

$$\mathrm{Op}_\tau(a)^\dagger = \mathrm{Op}_{1-\tau}(a^*). \tag{9.33}$$

In particular, we do not have $\widehat{A}_\tau^\dagger = \widehat{A}_\tau$ when the symbol a is real (unless $\tau = \frac{1}{2}$). Shubin pseudo-differential operators are therefore not suitable for physical quantization, where quantization should turn a real observable into a selfadjoint operator. From the physical point of view these operators are only useful intermediates for the definition of the physically consistent Born–Jordan quantization procedure we study in the next chapter.

9.3 Back to Shubin's Operators

The machinery developed above allows us to study in a rather neat and simple way the main properties of Shubin operators defined at the beginning of this chapter.

9.3.1 Harmonic Decomposition of τ-Operators

The Shubin pseudo-differential operators and the τ-cross-Wigner transform are related by the following formula, which is the analogue of formula (7.19) in Proposition 8 in Chap. 7:

Proposition 14 *Let $\psi, \phi \in \mathcal{S}(\mathbb{R}^n)$, $a \in \mathcal{S}'(\mathbb{R}^{2n})$, and $\tau \in \mathbb{R}$. Let $\widehat{A}_\tau = \operatorname{Op}_\tau(a)$. (i) We have*

$$\langle \widehat{A}_\tau \psi, \phi^* \rangle = \langle\langle a, \operatorname{Wig}_\tau(\psi, \phi) \rangle\rangle \tag{9.34}$$

where $\langle\langle \cdot, \cdot \rangle\rangle$ is the distributional bracket on $\mathbb{R}^{2n}$; in integral notation

$$\int \widehat{A}_\tau \psi(x)\phi^*(x)d^n x = \int a(z)\operatorname{Wig}_\tau(\psi, \phi)(z)d^{2n}z. \tag{9.35}$$

(ii) Similarly,

$$\int \widehat{A}_\tau \psi(x)\phi^*(x)d^n x = \int a_\sigma(z)\operatorname{Amb}_\tau(\psi, \phi)(-z)d^{2n}z \tag{9.36}$$

where $a_\sigma = F_\sigma a$ is the covariant symbol of $\widehat{A}_\tau$.

Proof (i) It suffices to assume that $a \in \mathcal{S}(\mathbb{R}^{2n})$. By definition of Wig_τ we have

$$\langle\langle a, \operatorname{Wig}_\tau(\psi, \phi) \rangle\rangle = \left(\tfrac{1}{2\pi\hbar}\right)^n \int e^{-\frac{i}{\hbar}py} a(x, p)\psi(x + \tau y) \times \phi^*(x - (1-\tau)y)d^n y d^n p d^n x;$$

setting $x' = x - (1-\tau)y$ and $y' = x + \tau y$ we get

$$\langle\langle a, \operatorname{Wig}_\tau(\psi, \phi) \rangle\rangle = \left(\tfrac{1}{2\pi\hbar}\right)^n \int e^{-\frac{i}{\hbar}p(x'-y')} \times a((1-\tau)x' + \tau y', p)\psi(y')\phi^*(x')d^n y' d^n p d^n x'$$

hence the equality (9.34) in view of definition (9.2) and (9.3) of the kernel of the Shubin operator $\widehat{A}_\tau = \operatorname{Op}_\tau(a)$ (or, using the integral definition (9.1) of that operator). (ii) Formula (9.36) follows from formula (9.35) using Plancherel's formula (6.23) and definition (9.25) of the τ-cross-ambiguity transform. ■

Formula (9.34) allows us to define $\widehat{A}_\tau\psi = \mathrm{Op}_\tau(a)\psi$ for arbitrary symbols $a \in \mathcal{S}'(\mathbb{R}^{2n})$ and $\psi \in \mathcal{S}(\mathbb{R}^n)$ in the same way as is done for Weyl pseudo-differential operators: choose $\phi \in \mathcal{S}(\mathbb{R}^n)$; then $\mathrm{Wig}_\tau(\psi, \phi) \in \mathcal{S}(\mathbb{R}^{2n})$ and the distributional bracket $\langle a, \mathrm{Wig}_\tau(\psi, \phi)\rangle$ is thus well-defined; by definition $\widehat{A}_\tau\psi$ is given by (9.34), and $\widehat{A}_\tau$ is a continuous operator $\mathcal{S}(\mathbb{R}^n) \longrightarrow \mathcal{S}'(\mathbb{R}^n)$.

The following consequence of Proposition 14, announced at the beginning of this chapter, justifies *a posteriori* the introduction of the τ-Grossmann–Royer and Heisenberg operators:

Corollary 15 *Let $\widehat{A}_\tau = \mathrm{Op}_\tau(a)$, $a \in S'(\mathbb{R}^{2n})$. We have*

$$\widehat{A}_\tau = \left(\tfrac{1}{\pi\hbar}\right)^n \int a(z)\widehat{T}_{\mathrm{GR},1-\tau}(z)d^{2n}z \tag{9.37}$$

$$\widehat{A}_\tau = \left(\tfrac{1}{2\pi\hbar}\right)^n \int a_\sigma(z)\widehat{T}_{1-\tau}(z)d^{2n}z \tag{9.38}$$

where a_σ is the symplectic Fourier transform of a.

Proof It is sufficient to prove these formulas for $a \in S(\mathbb{R}^{2n})$, the general case following by duality and continuity. Let ϕ and ψ be in $\mathcal{S}(\mathbb{R}^n)$. In view of formula (9.36) in Proposition 14 we have

$$\int \widehat{A}_\tau\psi(x)\phi^*(x)d^nx = \int a_\sigma(z)\mathrm{Amb}_\tau(\psi, \phi)(-z)d^{2n}z$$

hence, taking formula (9.27) in Proposition 11 into account,

$$\int \widehat{A}_\tau\psi(x)\phi^*(x)d^nx = \int a_\sigma(z)e^{-\frac{i}{\hbar}(2\tau-1)px}\mathrm{Amb}(\psi, \phi)(-z)d^{2n}z.$$

Since, by definition (7.11),

$$\mathrm{Amb}(\psi, \phi)(z) = \left(\tfrac{1}{2\pi\hbar}\right)^n \langle\widehat{T}(z)\phi|\psi\rangle$$

this equality can be written

$$\int \widehat{A}_\tau\psi(x)\phi^*(x)d^nx = \left(\tfrac{1}{2\pi\hbar}\right)^n \int a_\sigma(z_0)e^{-\frac{i}{\hbar}(2\tau-1)p_0x_0}\langle\widehat{T}(-z_0)\phi|\psi\rangle d^{2n}z_0;$$

Heisenberg operators being unitary we have

$$\langle\widehat{T}(-z_0)\phi|\psi\rangle = \langle\widehat{T}(z_0)^\dagger\phi|\psi\rangle = \langle\phi|\widehat{T}(z_0)\psi\rangle$$

and hence

$$\int \widehat{A}_\tau \psi(x)\phi^*(x)d^n x = \left(\tfrac{1}{2\pi\hbar}\right)^n \int a_\sigma(z_0)e^{-\frac{i}{\hbar}(2\tau-1)p_0x_0}\langle\phi|\widehat{T}(z_0)\psi\rangle d^{2n}z_0$$

$$= \left(\tfrac{1}{2\pi\hbar}\right)^n \int a_\sigma(z_0)e^{-\frac{i}{\hbar}(2\tau-1)p_0x_0}$$

$$\times\left(\int \phi^*(x)\widehat{T}(z_0)\psi(x)d^n x\right) d^{2n}z_0$$

which can be rewritten as

$$\int \widehat{A}_\tau \psi(x)\phi^*(x)d^n x = \left(\tfrac{1}{2\pi\hbar}\right)^n$$

$$\times \int\left(\int a_\sigma(z_0)e^{-\frac{i}{\hbar}(2\tau-1)p_0x_0}\widehat{T}(z_0)\psi(x)d^{2n}z_0\right)\phi^*(x)d^n x;$$

since ϕ is an arbitrary element of $\mathcal{S}(\mathbb{R}^n)$ we thus have

$$\widehat{A}_\tau \psi(x) = \int a_\sigma(z_0)e^{-\frac{i}{\hbar}(2\tau-1)p_0x_0}\widehat{T}(z_0)\psi(x)d^{2n}z_0$$

$$= \int a_\sigma(z_0)\widehat{T}_{1-\tau}(z_0)\psi(x)d^{2n}z_0$$

which proves (9.38). Formula (9.37) follows using Parseval's formula and the relation (9.15). ■

9.3.2 *Products, Transposes, Adjoints*

The τ-operators can be composed exactly in the same way as usual Weyl operators, replacing the symplectic form σ in (6.55) by the bilinear form σ_τ defined by (9.14):

Proposition 16 *Let $\widehat{A}_\tau = \mathrm{Op}_\tau(a)$ and $B_\tau = \mathrm{Op}_\tau(b)$ with $a, b \in \mathcal{S}(\mathbb{R}^{2n})$ (or, more generally $a \in \mathcal{S}'(\mathbb{R}^{2n})$ and $b \in \mathcal{S}(\mathbb{R}^{2n})$). We have*

$$\widehat{A}_\tau B_\tau = C_\tau = \mathrm{Op}_\tau(c)$$

where the symplectic Fourier transform of the symbol c is given by

$$c_\sigma(z) = \left(\tfrac{1}{2\pi\hbar}\right)^n \int e^{\frac{i}{2\hbar}\sigma_{1-\tau}(z-z',z')}a_\sigma(z-z')b_\sigma(z')d^{2n}z' \tag{9.39}$$

where σ_τ is a bilinear form given by

$$\sigma_{1-\tau}(z_0, z_1) = 2\tau p_0 x_1 - 2(1-\tau) p_1 x_0. \tag{9.40}$$

(In formula (9.39) the integral should be viewed as a distributional bracket when $a \in \mathcal{S}'(\mathbb{R}^{2n})$ and $b \in \mathcal{S}(\mathbb{R}^{2n})$.)

Proof Writing

$$\widehat{A}_\tau = \left(\tfrac{1}{2\pi\hbar}\right)^n \int a_\sigma(z)\widehat{T}_{1-\tau}(z)d^{2n}z \tag{9.41}$$

$$B_\tau = \left(\tfrac{1}{2\pi\hbar}\right)^n \int b_\sigma(z)\widehat{T}_{1-\tau}(z)d^{2n}z \tag{9.42}$$

we have

$$\widehat{A}_\tau B_\tau = \int a_\sigma(z_0)b_\sigma(z_1)\widehat{T}_{1-\tau}(z_0)\widehat{T}_{1-\tau}(z_1)d^{2n}z_0 d^{2n}z_1$$

and hence, using formula (9.13),

$$\widehat{A}_\tau B_\tau = \int e^{\frac{i}{\hbar}\sigma_{1-\tau}(z_0,z_1)} a_\sigma(z_0)b_\sigma(z_1)\widehat{T}_{1-\tau}(z_0+z_1)d^{2n}z_0 d^{2n}z_1.$$

Making the change of variables $z = z_0+z_1$, $z' = z_1$, the composition formula (9.39) follows. ■

The transpose of an operator $A : \mathcal{S}(\mathbb{R}^n) \longrightarrow \mathcal{S}'(\mathbb{R}^n)$ is defined as follows: it is the unique operator $A^T : \mathcal{S}(\mathbb{R}^n) \longrightarrow \mathcal{S}'(\mathbb{R}^n)$ such that $\langle A\psi, \phi\rangle = \langle A^T\phi, \psi\rangle$ *for all* $\psi, \phi \in \mathcal{S}(\mathbb{R}^n)$. The formal adjoint of $A^\dagger$ of A is defined as usual by $\langle A\psi|\phi\rangle = \langle\psi|A^\dagger\phi\rangle$; equivalently:

$$A^\dagger\psi = (A^T\psi^*)^* \textit{ for all } \psi \in \mathcal{S}(\mathbb{R}^n). \tag{9.43}$$

Proposition 17 *Let $\widehat{A}_\tau = \operatorname{Op}_\tau(a)$. We have (i) $\widehat{A}_\tau^T = \operatorname{Op}_{1-\tau}(a^T)$ where*

$$a^T(x,p) = a(x,-p). \tag{9.44}$$

(ii) We have

$$\widehat{A}_\tau^\dagger = \operatorname{Op}_{1-\tau}(a^*) \tag{9.45}$$

Proof Assume that $a \in \mathcal{S}(\mathbb{R}^{2n})$. Using the integral representation (9.1) of $\widehat{A}_\tau$ we have

$$\langle\widehat{A}_\tau\psi, \phi\rangle = \left(\tfrac{1}{2\pi\hbar}\right)^n \int e^{\frac{i}{\hbar}p(x-y)} a((1-\tau)x+\tau y, p)\psi(y)\phi(x)d^n y d^{2n}z$$

$$\langle\widehat{A}_\tau^T\phi, \psi\rangle = \left(\tfrac{1}{2\pi\hbar}\right)^n \int e^{\frac{i}{\hbar}p(x-y)} a((1-\tau)x+\tau y, p)\phi(y)\psi(x)d^n y d^{2n}z.$$

Formula (9.44) follows, swapping the variables x and y in the second equality. The general case follows by continuity. (Alternatively one can use the harmonic decomposition (9.37) of $\widehat{A}_{T}$.) Formula (9.45) follows, using (9.43). ■

References

1. P. Boggiatto, G. De Donno, A. Oliaro, Time-frequency representations of Wigner type and pseudo-differential operators. Trans. Amer. Math. Soc. **362**(9), 4955–4981 (2010)
2. P. Boggiatto, B.K. Cuong, G. De Donno, A. Oliaro, Weighted integrals of Wigner representations. J. Pseudo-Diff. Oper. Appl. (2010)
3. M. de Gosson, *Symplectic Methods in Harmonic Analysis and in Mathematical Physics*, vol. 7 (Springer Science & Business Media, 2011)
4. K. Gröchenig, *Foundations of Time-Frequency Analysis* (Birkhäuser, Boston, 2000)
5. M.A. Shubin, *Pseudodifferential Operators and Spectral Theory* (Springer, 1987) (original Russian edition in Nauka, Moskva, 1978)
6. J. Toft, Multiplication properties in pseudo-differential calculus with small regularity on the symbols. J. Pseudo-Diff. Oper. Appl. **1**, 101–138 (2010)

Chapter 10
Born–Jordan Pseudo-Differential Operators

In Chap. 7 we defined Born–Jordan quantization by using the properties of the Cohen class. In this chapter we use the theory of Shubin operators and define the Born–Jordan operator $\widehat{A}_{\mathrm{BJ}}$ with symbol a as being the average over the interval $[0, 1]$ of the τ-operators $\widehat{A}_\tau$ with the same symbol. This definition is consistent with the observations made in our discussion of the monomial case in Chap. 3 where we showed that the Born–Jordan quantization of a monomial can be obtained by a similar averaging procedure. We will prove that this alternative definition leads to the same notion of operator as formerly; this will require some work relying on previous results due to Boggiatto and his collaborators [1–3], who addressed the problem of how to damp unwanted interference effects in time-frequency analysis.

10.1 Born–Jordan Pseudo-Differential Operators

We now study Born–Jordan operators using the Shubin pseudo-differential calculus developed in the last chapter.

10.1.1 Definition and Justification

Everything in this chapter stems from the following definition:

Definition 1 Let $a \in \mathcal{S}'(\mathbb{R}^{2n})$ be an arbitrary symbol. The Born–Jordan operator $\widehat{A}_{\mathrm{BJ}} = \mathrm{Op}_{\mathrm{BJ}}(a)$ is defined for $\psi \in \mathcal{S}(\mathbb{R}^n)$ by:

$$\widehat{A}_{\mathrm{BJ}}\psi(x) = \int_0^1 \widehat{A}_\tau \psi(x) d\tau \tag{10.1}$$

where $\widehat{A}_\tau = \mathrm{Op}_\tau(a)$ is the τ-operator with symbol a.

M.A. de Gosson, *Born–Jordan Quantization*, Fundamental Theories of Physics 182, DOI 10.1007/978-3-319-27902-2_10

It immediately follows from this definition that, as in the Weyl case, the formal adjoint of $\widehat{A}_{\mathrm{BJ}} = \mathrm{Op}_{\mathrm{BJ}}(a)$ is $\widehat{A}^{\dagger}_{\mathrm{BJ}} = \mathrm{Op}_{\mathrm{BJ}}(a^*)$: we have

$$\widehat{A}^{\dagger}_{\mathrm{BJ}} = \int_0^1 \mathrm{Op}_{1-\tau}(a^*)d\tau = \mathrm{Op}_{\mathrm{BJ}}(a^*).$$

In particular, the Born–Jordan operator $\widehat{A}_{\mathrm{BJ}}$ is formally self-adjoint if and only if its symbol is real.

To justify this definition (and terminology) we have to show that one recovers the Born–Jordan quantization formulas for polynomials discussed in Chap. 3. That is, we have to show that the definition (10.1) implies that we have

$$\mathrm{Op}_{\mathrm{BJ}}(x^r p^s) = \frac{1}{s+1}\sum_{k=0}^{s} \widehat{p}^{s-k}\widehat{x}^r\widehat{p}^k. \tag{10.2}$$

In Proposition 1 of Chap. 9 we showed that the τ-operator associated with the monomial $x^r p^s$ is given by any of the two equivalent expressions

$$\mathrm{Op}_{\tau}(x^r p^s) = \sum_{k=0}^{s} \binom{s}{k} (1-\tau)^k \tau^{s-k} \widehat{p}^k\widehat{x}^r\widehat{p}^{s-k} \tag{10.3}$$

$$\mathrm{Op}_{\tau}(x^r p^s) = \sum_{k=0}^{r} \binom{r}{k} \tau^k (1-\tau)^{r-k} \widehat{x}^k\widehat{p}^s\widehat{x}^{r-k}. \tag{10.4}$$

Integrating in τ from 0 to 1 the equality (10.3) yields

$$\mathrm{Op}_{\mathrm{BJ}}(x^r p^s) = \sum_{k=0}^{s} \binom{s}{k} B(s-k+1, k+1)\widehat{p}^k\widehat{x}^r\widehat{p}^{s-k}$$

where

$$B(u,v) = \int_0^1 t^{u-1}(1-t)^{v-1}dt = \frac{\Gamma(u)\Gamma(v)}{\Gamma(u+v)}$$

is Euler's beta function (Jeffreys [7]). Since

$$B(k+1, s-k+1) = \frac{k!(s-k)!}{(s+1)!}$$

we have

$$\operatorname{Op_{BJ}}(x^r p^s) = \frac{1}{s+1}\sum_{k=0}^{r} \widehat{p}^k \widehat{x}^r \widehat{p}^{s-k}$$

which is the same thing as (10.2). Using formula (10.4) the same argument would yield the alternative equality

$$\operatorname{Op_{BJ}}(x^r p^s) = \frac{1}{r+1}\sum_{k=0}^{r} \widehat{x}^k \widehat{p}^s \widehat{x}^{r-k}. \tag{10.5}$$

An immediate consequence of the discussion above is that the Weyl and Born–Jordan quantizations are identical for all quadratic Hamiltonians. Suppose in fact that

$$H(z) = \sum_j \alpha_j p_j^2 + \beta_j x_j^2 + 2\gamma_j p_j x_j.$$

We immediately see that $\operatorname{Op_{BJ}}(H) = \operatorname{Op_W}(H)$ when the γ_j are all zero; when there are cross-terms $x_j p_j$ the claim follows using formula (10.2) (or (10.5)) with $r = s = 1$; this shows that the Born–Jordan quantization of $x_j p_j$ is $\frac{1}{2}(\widehat{x_j}\widehat{p_j} + \widehat{p_j}\widehat{x_j})$ which is the same result as that obtained using Weyl quantization (cf. formula (2.10)). In both cases, the corresponding operator is thus formally given by

$$\widehat{H} = \frac{1}{2}(\widehat{x}, \widehat{p}) M (\widehat{x}, \widehat{p})^T$$

where M is the Hessian matrix of H. The quadratic case is important in the context of the metaplectic group and will be discussed in Chap. 12: it shows that the "lift" of a linear Hamiltonian flow to the metaplectic group leads to the same Schrödinger equation, whether one uses the Weyl or the Born–Jordan scheme.

We will see later in this chapter that the usual "physical" Hamiltonians also have the same quantizations in both the Weyl and Born–Jordan schemes, even when a magnetic field is present.

10.1.2 Born–Jordan and Weyl Operators

The results above are of course not the end of the game, because we still have to prove that the rule (10.1) leads to the same operators as those defined in Chap. 7 using the Born–Jordan kernel.

Recall that the sinc function was defined in Sect. 8.1.

Proposition 2 *(i) Let* $a \in \mathcal{S}'(\mathbb{R}^{2n})$ *and* $\psi \in \mathcal{S}(\mathbb{R}^n)$*. The Born–Jordan operator* $\widehat{A}_{\mathrm{BJ}} = \mathrm{Op}_{\mathrm{BJ}}(a)$ *is given by*

$$\widehat{A}_{\mathrm{BJ}}\psi = \left(\tfrac{1}{2\pi\hbar}\right)^n \int a_\sigma(z)\chi_{\mathrm{BJ}}(z)\widehat{T}(z)\psi d^{2n}z \tag{10.6}$$

where the function χ_{BJ} *is defined by*

$$\chi_{\mathrm{BJ}}(z) = \mathrm{sinc}(px/2\hbar). \tag{10.7}$$

(ii) The covariant Weyl symbol $a_{\mathrm{BJ},\sigma}$ *of* $\widehat{A}_{\mathrm{BJ}}$ *is given by the explicit formula*

$$a_{\mathrm{BJ},\sigma}(z) = a_\sigma(z)\chi_{\mathrm{BJ}}(z). \tag{10.8}$$

(iii) In particular the operator $\widehat{A}_{\mathrm{BJ}}$ *is a continuous operator* $\mathcal{S}(\mathbb{R}^n) \longrightarrow \mathcal{S}'(\mathbb{R}^n)$ *for every* $a \in \mathcal{S}'(\mathbb{R}^{2n})$*.*

Proof The statement (ii) immediately follows from formula (10.6) taking the harmonic representation (6.40) of Weyl operators into account. The proof of formula (10.6) goes as follows (cf. [6], Proposition 11). Recall (formula (9.38) in Corollary 1 in Chap. 9) that we have

$$\widehat{A}_\tau\psi = \left(\tfrac{1}{2\pi\hbar}\right)^n \int a_\sigma(z)\widehat{T}_{1-\tau}(z)\psi d^{2n}z.$$

Integrating both sides of the equality with respect to $\tau \in [0,1]$ we get

$$\widehat{A}_{\mathrm{BJ}} = \left(\tfrac{1}{2\pi\hbar}\right)^n \int a_\sigma(z)\left(\int_0^1 \widehat{T}_{1-\tau}(z)d\tau\right) d^{2n}z.$$

Using (formula 9.10), we have for $px \neq 0$,

$$\begin{aligned}\int_0^1 \widehat{T}_\tau(z)d\tau &= \left(\int_0^1 e^{\frac{i}{2\hbar}(2\tau-1)px}d\tau\right)\widehat{T}(z)\\ &= \frac{\sin(px/2\hbar)}{px/2\hbar}\widehat{T}(z)\\ &= \mathrm{sinc}\left(\frac{px}{2\hbar}\right)\widehat{T}(z)\end{aligned}$$

hence formula (10.6), noting that the equality holds by continuity for $px = 0$. (iii) The statement follows from the fact that $a_{\mathrm{BJ},\sigma} \in \mathcal{S}'(\mathbb{R}^n)$ since $\partial_z^\alpha\chi_{\mathrm{BJ}} \in C^\infty(\mathbb{R}^{2n}) \cap L^\infty(\mathbb{R}^{2n})$ for all $\alpha \in \mathbb{N}^{2n}$. ■

The proof of the following composition result immediately follows from the characterization (10.8) above using Proposition 18 in Chap. 6 (ii):

Proposition 3 *Let $\widehat{A}_{\mathrm{BJ}} = \mathrm{Op}_{\mathrm{BJ}}(a)$ and $\widehat{B}_{\mathrm{BJ}} = \mathrm{Op}_{\mathrm{BJ}}(b)$ be two Born–Jordan pseudo-differential operators; we suppose that $\widehat{C} = \widehat{A}_{\mathrm{BJ}}\widehat{B}_{\mathrm{BJ}}$ exists as an operator $\mathcal{S}(\mathbb{R}^n) \longrightarrow \mathcal{S}'(\mathbb{R}^n)$. (i) The covariant symbol of the Weyl operator $\widehat{C} = \mathrm{Op}_{\mathrm{W}}(c)$ is given by the formula*

$$c_\sigma^{\mathrm{W}}(z) = \int e^{\frac{i}{2\hbar}\sigma(z,z')} a_\sigma(z-z')b(z')\chi_{\mathrm{BJ}}(z-z')\chi_{\mathrm{BJ}}(z')d^{2n}z' \tag{10.9}$$

where χ_{BJ} is defined by (10.7)*. (ii) If we can factorize c_σ^{W} as $c_\sigma^{\mathrm{W}}(z) = \chi(z)\chi_{\mathrm{BJ}}(z)$ where $\chi \in \mathcal{S}'(\mathbb{R}^n)$ then $\chi = c_\sigma$ with $\widehat{C} = \mathrm{Op}^{\mathrm{BJ}}(c)$.*

Note that neither χ nor c are uniquely defined by the relation $c_\sigma^{\mathrm{W}}(z) = \chi(z)\chi_{\mathrm{BJ}}(z)$ since $\chi_{\mathrm{BJ}}(z) = 0$ for infinitely many values of z.

10.2 The Born–Jordan Transform Revisited

In [1, 2] Boggiatto et al. consider the average of the τ-cross-Wigner transforms $W_\tau(\psi, \phi)$ over the interval $[0, 1]$ that is

$$Q(\psi, \phi)(z) = \int_0^1 \mathrm{Wig}_\tau(\psi, \phi)(z)dt. \tag{10.10}$$

It immediately follows from (9.28) that the marginal properties hold for the distribution $Q\psi = Q(\psi, \phi)$: if $\psi \in L^1(\mathbb{R}^n) \cap L^2(\mathbb{R}^n)$ and $\widehat{\psi} \in L^1(\mathbb{R}^n)$ then

$$\int \mathrm{Wig}_{\mathrm{BJ}}\psi(z)d^n p = |\psi(x)|^2, \int \mathrm{Wig}_{\mathrm{BJ}}\psi(z)d^n x = |\widehat{\psi}(p)|^2. \tag{10.11}$$

In fact, it is easy to verify, using Proposition 11 in Chap. 7 that Q belongs to the Cohen class. We have in fact the following essential result which shows that the distribution of Boggiatto et al. is just the Born–Jordan distribution earlier defined in Chap. 8 in Definition 1):

Proposition 4 *The quasi-distribution associated to* (10.10) *by*

$$Q\psi(z) = \int_0^1 \mathrm{Wig}_\tau\psi(z)d\tau$$

is the Born–Jordan quasi-distribution (8.4), *that is*

$$Q\psi = \mathrm{Wig}_{\mathrm{BJ}}\psi * \theta_{\mathrm{BJ}} \tag{10.12}$$

where θ_{BJ} *is the Born–Jordan kernel.*

Proof See Boggiatto et al. [1]. ∎

As expected from the Weyl and Shubin cases, the Born–Jordan operators can be expressed in terms of their symbol and $\mathrm{Wig}_{\mathrm{BJ}}(\psi, \phi)$; we have already proved this in Chap. 7, but here is an independent alternative proof using the approach outlined above:

Proposition 5 *Let* $a \in \mathcal{S}'(\mathbb{R}^{2n})$ *be a symbol. The operator* $\widehat{A}_{\mathrm{BJ}} = \mathrm{Op}_{\mathrm{BJ}}(a)$ *and the cross-Born–Jordan transform* $\mathrm{Wig}_{\mathrm{BJ}}(\psi, \phi)$ *are related by the formula*

$$\langle \widehat{A}_{\mathrm{BJ}}\psi, \phi^* \rangle = \langle\langle a, \mathrm{Wig}_{\mathrm{BJ}}(\psi, \phi)\rangle\rangle \tag{10.13}$$

for all $(\psi, \phi) \in \mathcal{S}(\mathbb{R}^n) \times \mathcal{S}(\mathbb{R}^n)$*; in integral notation*

$$\int \widehat{A}_{\mathrm{BJ}}\psi(x)\phi^*(x)d^n x = \int a(z)\mathrm{Wig}_{\mathrm{BJ}}(\psi, \phi)(z)d^{2n}z. \tag{10.14}$$

Proof In view of formula (9.34) we have

$$\langle \widehat{A}_\tau\psi, \phi^* \rangle = \langle\langle a, \mathrm{Wig}_\tau(\psi, \phi)\rangle\rangle;$$

integrating this equality from 0 to 1 with respect to the variable τ yields (10.13). ∎

10.2.1 Born–Jordan Versus Weyl

Born–Jordan and Weyl quantization are identical for "physical" Hamiltonians of the type "*kinetic energy* plus *potential*". Suppose in fact that

$$H(z) = \sum_{j=1}^{n} \frac{1}{2m_j} p_j^2 + V(x) \tag{10.15}$$

then $\widehat{H} = \mathrm{Op}_{\mathrm{BJ}}(H)$ is given by

$$\widehat{H} = -\sum_{j=1}^{n} \frac{\hbar^2}{2m_j} \frac{\partial^2}{\partial x_j^2} + V(x). \tag{10.16}$$

This can be seen by noting that $\mathrm{Op_{BJ}}(p_j^2) = -\hbar^2\partial^2/\partial x_j^2$ taking $r = 0$ and $s = 2$ in formula (10.5) and then using definition (9.1):

$$\begin{aligned}\mathrm{Op}_\tau(V)\psi(x) &= \left(\tfrac{1}{2\pi\hbar}\right)^n \int e^{\frac{i}{\hbar}p(x-y)} V(\tau x + (1-\tau)y)\psi(y)d^n y d^n p \\ &= \int V(\tau x + (1-\tau)y)\psi(y)\delta(x-y)d^n y \\ &= V(x)\psi(x);\end{aligned}$$

integrating in τ from 0 to 1 yields $\mathrm{Op_{BJ}}(V)\psi = V\psi$ and hence (10.16).

More generally the Born–Jordan and Weyl quantizations of the magnetic Hamiltonian

$$H(z,t) = \sum_{j=1}^{n} \frac{1}{2m_j}\left(p_j - A_j(x,t)\right)^2 + V(x,t) \tag{10.17}$$

also coincide. To show this we need the following intermediary result:

Lemma 6 *Let $A : \mathbb{R}^n \times \mathbb{R}_t \longrightarrow \mathbb{R}$ be a smooth function. Then*

$$\mathrm{Op_{BJ}}(p_j\mathcal{A})\psi = \mathrm{Op_W}(p_j\mathcal{A})\psi = -\frac{i\hbar}{2}\left[\frac{\partial}{\partial x}(\mathcal{A}\psi) + \mathcal{A}\frac{\partial}{\partial x}\psi\right]. \tag{10.18}$$

Proof It is sufficient to assume $n = 1$. Using definition (9.1) of $\widehat{A}_\tau = \mathrm{Op}_\tau(a)$ we have

$$\begin{aligned}\mathrm{Op}_\tau(p\mathcal{A})\psi(x) &= \tfrac{1}{2\pi\hbar}\int e^{\frac{i}{\hbar}p(x-y)} p\mathcal{A}(\tau x + (1-\tau)y, t)\psi(y)dydp \\ &= \int_{-\infty}^{\infty}\left[\tfrac{1}{2\pi\hbar}\int_{-\infty}^{\infty} e^{\frac{i}{\hbar}p(x-y)} pdp\right]\mathcal{A}(\tau x + (1-\tau)y, t)\psi(y)dy.\end{aligned}$$

In view of formula (9.7) the expression between the square brackets is $-i\hbar\delta'(x-y)$ hence

$$\begin{aligned}\mathrm{Op}_\tau(p\mathcal{A})\psi(x) &= -i\hbar\int_{-\infty}^{\infty}\delta'(x-y)\mathcal{A}(\tau x + (1-\tau)y, t)\psi(y)dy \\ &= -i\hbar\int_{-\infty}^{\infty}\delta(x-y)\tfrac{\partial}{\partial y}\left[\mathcal{A}(\tau x + (1-\tau)y, t)\psi(y)\right]dy \\ &= -i\hbar\left[(1-\tau)\tfrac{\partial}{\partial x}(\mathcal{A}\psi) + \tau\mathcal{A}\tfrac{\partial}{\partial x}\psi\right].\end{aligned}$$

Formula (10.18) follows setting in the Weyl case $\tau = \frac{1}{2}$ and integrating from 0 to 1 in the Born–Jordan case. ■

It follows that both Weyl and Born–Jordan quantizations of a (time-dependent) magnetic Hamiltonian (10.17) are the same. In fact, expanding the terms $(p_j - \mathcal{A}_j(x,t))^2$ we get

$$H = \sum_{j=1}^{n} \frac{1}{2m_j} p_j^2 - \sum_{j=1}^{n} \frac{1}{m_j} p_j \mathcal{A}_j + \sum_{j=1 j}^{n} \mathcal{A}_j^2 + V.$$

We have seen above that the terms p_j^2 and $\mathcal{A}_j^2 + V$ have identical quantizations; in view of formula (10.18) this also true of the cross-terms $p_j A_j$, leading in both cases to the expression

$$\widehat{H} = \sum_{j=1}^{n} \frac{1}{2m_j} \left(-i\hbar \frac{\partial}{\partial x_j} - \mathcal{A}_j^2(x,t) \right)^2 + V(x,t) \tag{10.19}$$

well-known from standard quantum mechanics.

10.3 Tensor Products of Observables

Let b and c be classical observables, defined on, respectively $\mathbb{R}^{2n_1}$ and $\mathbb{R}^{2n_2}$. Writing $n = n_1 + n_2$ their tensor product $b \otimes c$ is the observable on $\mathbb{R}^{2n} \equiv \mathbb{R}^{2n_1} \times \mathbb{R}^{2n_2}$ defined by

$$(b \otimes c)(z_1, z_2) = b(z_1)c(z_2)$$

(it is indeed an observable, because tensor products of tempered distributions are themselves tempered distributions).

10.3.1 Weyl Operators

We now address the following question: for which quantizations Op do we have $\mathrm{Op}(b \otimes c) = \mathrm{Op}(b) \otimes \mathrm{Op}(c)$? Let us prove that this is the case for Weyl quantization:

Proposition 7 *Let $b \in \mathcal{S}'(\mathbb{R}^{2n_1})$ and $c \in \mathcal{S}'(\mathbb{R}^{2n_2})$. We have $b \otimes c \in \mathcal{S}'(\mathbb{R}^{2n})$ with $n = n_1 + n_2$ and*

$$\mathrm{Op_W}(b \otimes c) = \mathrm{Op_W}(b) \otimes \mathrm{Op_W}(c). \tag{10.20}$$

Proof Using the harmonic representation (6.40) of Weyl operators in Sect. 6.3, the operator $\widehat{A} = \mathrm{Op_W}(b \otimes c)$ is given by

$$\widehat{A} = \left(\tfrac{1}{2\pi\hbar}\right)^n \int (b \otimes c)_\sigma(z_0)\widehat{T}(z_0)d^{2n}z_0;$$

now the symplectic Fourier transform respects (as does the usual Fourier transform) tensor products: $(b \otimes c)_\sigma = b_\sigma \otimes c_\sigma$ so we have

$$\widehat{A} = \left(\tfrac{1}{2\pi\hbar}\right)^n \int (b_\sigma \otimes c_\sigma)(z_0)\widehat{T}(z_0)d^{2n}z_0. \tag{10.21}$$

Split now the phase space variable z in two two components $z_1 \in \mathbb{R}^{2n_1}$ and $z_2 \in \mathbb{R}^{2n_2}$ so that $z = (z_1, z_2)$, and denote by σ_1 and σ_2 the symplectic forms on $\mathbb{R}^{2n_1}$ and $\mathbb{R}^{2n_2}$, respectively. We clearly have $\sigma = \sigma_1 \oplus \sigma_2$, that is

$$\sigma(z, z') = \sigma_1(z_1, z_1') + \sigma_2(z_2, z_2');$$

on the operator level this relation becomes, setting $\widehat{z} = (\widehat{z}_1, \widehat{z}_2)$ and $z' = (z_1', z_2')$,

$$\sigma(\widehat{z}, z') = \sigma_1(\widehat{z}_1, z_1') + \sigma_2(\widehat{z}_2, z_2').$$

It follows that the Heisenberg operator $\widehat{T}(z_0) = e^{-\frac{i}{\hbar}\sigma(\hat{z}, z_0)}$ can be written as a tensor product

$$\widehat{T}(z_0) = \widehat{T}(z_{0,1}) \otimes \widehat{T}(z_{0,2}) \tag{10.22}$$

where $z_0 = (z_{0,1}, z_{0,2})$. Insertion in (10.21) then yields

$$\widehat{A} = \left(\tfrac{1}{2\pi\hbar}\right)^n \int b_\sigma(z_{0,1})c_\sigma(z_{0,2})\widehat{T}(z_{0,1}) \otimes \widehat{T}(z_{0,2})d^{2n_1}z_{0,1}d^{2n_2}z_{0,2}$$

that is $\widehat{A} = \widehat{B} \otimes \widehat{C}$ with

$$\widehat{B} = \left(\tfrac{1}{2\pi\hbar}\right)^{n_1} \int b_\sigma(z_{0,1})\widehat{T}(z_{0,1})d^{2n_1}z_{0,1}$$
$$\widehat{C} = \left(\tfrac{1}{2\pi\hbar}\right)^{n_2} \int c_\sigma(z_{02})\widehat{T}(z_{0,2})d^{2n_2}z_{0,2}$$

where $\widehat{T}(z_{0,1})$ and $\widehat{T}(z_{0,2})$ are Heisenberg operators on $\mathbb{R}^{2n_1}$ and $\mathbb{R}^{2n_2}$, respectively. Remarking that $\widehat{B} = \mathrm{Op_W}(b)$ and $\widehat{C} = \mathrm{Op_W}(c)$ concludes the proof. ■

A similar result follows for the cross-ambiguity and Wigner transforms:

Proposition 8 *Let $\psi_1, \phi_1 \in \mathcal{S}(\mathbb{R}^{n_1})$ and $\psi_2, \phi_2 \in \mathcal{S}(\mathbb{R}^{n_2})$; then $\psi_1 \otimes \psi_2 \in \mathcal{S}(\mathbb{R}^n)$ where $n = n_1 + n_2$, and we have*

$$\mathrm{Wig}(\psi_1 \otimes \psi_2, \phi_1 \otimes \phi_2) = \mathrm{Wig}_1(\psi_1, \phi_1) + \mathrm{Wig}_2(\psi_2, \phi_2) \tag{10.23}$$
$$\mathrm{Amb}(\psi_1 \otimes \psi_2, \phi_1 \otimes \phi_2) = \mathrm{Amb}_1(\psi_1, \phi_1) + \mathrm{Amb}_2(\psi_2, \phi_2) \tag{10.24}$$

where Wig_1 *and* Wig_2 *(resp.* Amb_1 *and* Amb_2*) are the cross-Wigner (resp. cross-ambiguity) transforms on* $\mathbb{R}^{n_1}$ *and* $\mathbb{R}^{n_2}$*.*

Proof Formula (10.24) immediately follows from the splitting (10.22) of the Heisenberg operator using definition (7.11) of the cross-ambiguity transform. Formula (10.23) is proven in a similar manner using definition (7.3) of the cross-Wigner transform, noting that the Grossmann–Royer operator can be split as

$$\widehat{T}_{\mathrm{GR}}(z) = \widehat{T}_{\mathrm{GR}}(z_1) \otimes \widehat{T}_{\mathrm{GR}}(z_2). \tag{10.25}$$

■

10.3.2 The Born–Jordan Case

The situation is very different for Born–Jordan operators. It may very well happen (as in the example in the previous subsection) that

$$\mathrm{Op}_{\mathrm{BJ}}(b \otimes c) \neq \mathrm{Op}_{\mathrm{BJ}}(b) \otimes \mathrm{Op}_{\mathrm{BJ}}(c). \tag{10.26}$$

To see this, we recall that a Born–Jordan operator $\widehat{A}_{\mathrm{BJ}} = \mathrm{Op}_{\mathrm{BJ}}(a)$ can always be written

$$\widehat{A}_{\mathrm{BJ}} = \left(\tfrac{1}{2\pi\hbar}\right)^n \int a_\sigma(z)\,\mathrm{sinc}\left(\tfrac{px}{2\hbar}\right)\widehat{T}(z)d^{2n}z \tag{10.27}$$

where

$$\mathrm{sinc}\left(\frac{px}{2\hbar}\right) = \frac{\sin(px/2\hbar)}{px/2\hbar}.$$

It follows that if $b \in \mathcal{S}'(\mathbb{R}^{2n_1})$ and $c \in \mathcal{S}'(\mathbb{R}^{2n_2})$ then the integrand in

$$\mathrm{Op}_{\mathrm{BJ}}(b \otimes c) = \left(\tfrac{1}{2\pi\hbar}\right)^n \int b_\sigma(z_1)c_\sigma(z_2)\,\mathrm{sinc}\left(\tfrac{p_1x_1+p_2x_2}{2\hbar}\right)\widehat{T}(z)d^{2n_1}z_1 d^{2n_2}z_2$$

cannot in general be split into the product of a function depending only on z_1 and one depending only on z_2, as in the Weyl case, because we have

$$\mathrm{sinc}\left(\frac{p_1x_1+p_2x_2}{2\hbar}\right) \neq \mathrm{sinc}\left(\frac{p_1x_1}{2\hbar}\right)\mathrm{sinc}\left(\frac{p_2x_2}{2\hbar}\right).$$

This is of course not a peculiarity of the Born–Jordan quantization per se; the same phenomenon happens when one deals with an arbitrary Cohen kernel which fails to split under tensor products, i.e. such that $\theta(x,p) \neq \theta_1(x_1,p_1)\theta_2(x_2,p_2)$. This also shows how simple Weyl quantization is compared with all other schemes.

10.3.3 Illustration: The "Angular Momentum Dilemma"

There is a little annoying practical fact which is being episodically discussed and commented upon in the literature: the Weyl correspondence fails to transform the square of the classical angular momentum to its accepted quantum analogue. In fact, the Weyl quantization of the classical angular-momentum-squared is not just the quantum angular momentum squared operator, but it further contains a constant term $\frac{3}{2}\hbar^2$ (formula (10.32)). This extra term is actually physically significant, since it accounts for the nonvanishing angular momentum of the ground-state Bohr orbit in the hydrogen atom (see Dahl and Schleich [4]). This vexing fact already puzzled Pauling in his textbook [8], and has been later taken up by Shewell [9], Dahl and Springborg [5], and many others. We now come back to the angular momentum dilemma, which we solve in two different ways using Born–Jordan quantization. This dilemma can be stated as follows: how does one bring the fact that the orbital angular momentum $\ell = \mathbf{r} \times \mathbf{p}$ in a Bohr orbit of the $1s$ state of the hydrogen atom is $\hbar$ into accordance with the fact that the same angular momentum in conventional quantum mechanics based on the Schrödinger picture is zero? In fact, writing $\mathbf{r} = (x_1, x_2, x_3)$ and $\mathbf{p} = (p_1, p_2, p_3)$ we have

$$\ell = (x_2 p_3 - x_3 p_2, x_3 p_1 - x_1 p_3, x_1 p_2 - x_2 p_1) \tag{10.28}$$

as the classical angular momentum vector. Consider now the square

$$\ell_3^2 = x_1^2 p_2^2 + x_2^2 p_1^2 - 2x_1 p_1 x_2 p_2 \tag{10.29}$$

of the component $\ell_3 = x_1 p_2 - x_2 p_1$; its Weyl quantization is

$$(\widehat{\ell_3^2})_{\mathrm{W}} = \widehat{x}_1^2 \widehat{p}_2^2 + \widehat{x}_2^2 \widehat{p}_1^2 - \tfrac{1}{2}(\widehat{x}_1 \widehat{p}_1 + \widehat{p}_1 \widehat{x}_1)(\widehat{x}_2 \widehat{p}_2 + \widehat{p}_2 \widehat{x}_2). \tag{10.30}$$

It turns out that this is in contradiction with what quantum mechanics in the Schrödinger picture predicts, and which is

$$(\widehat{\ell_3^2})_{\mathrm{QM}} = \widehat{x}_1^2 \widehat{p}_2^2 + \widehat{x}_2^2 \widehat{p}_1^2 - (\widehat{x}_1 \widehat{p}_1 \widehat{p}_2 \widehat{x}_2 + \widehat{p}_1 \widehat{x}_1 \widehat{x}_2 \widehat{p}_2). \tag{10.31}$$

(This is obtained by formally expanding the square $(\widehat{x}_1 \widehat{p}_2 - \widehat{x}_2 \widehat{p}_1)^2$.) Using the commutation relations $[\widehat{x}_1, \widehat{p}_1] = i\hbar$ twice, one shows that the expressions (10.30) and (10.31) differ by the quantity $\frac{1}{2}\hbar^2$, and this leads to an overall difference of $\frac{3}{2}\hbar^2$ between these quantizations of ℓ^2:

$$(\widehat{\ell^2})_{\mathrm{W}} = (\widehat{\ell^2})_{\mathrm{QM}} + \tfrac{3}{2}\hbar^2 \tag{10.32}$$

(cf. Shewell [9], formula (4.10); for a generalization to higher dimensions see Dahl and Schleich [4]).

Let us show that this contradiction disappears if we use the Born–Jordan quantization of the classical quantity ℓ^2 in place of Weyl quantization. The two terms $x_1^2 p_2^2$ and $x_2^2 p_1^2$ immediately yield $\widehat{x_1^2}\widehat{p_2^2}$ and $\widehat{x_2^2}\widehat{p_1^2}$ (as they would in any quantization scheme), so let us focus on the cross-term $a(z) = 2x_1 p_1 x_2 p_2$ (we are writing here $z = (x_1, x_2, p_1, p_2)$). Using the standard formula giving the Fourier transform of a monomial we get

$$a_\sigma(z) = 2\hbar^4 (2\pi\hbar)^2 \delta'(z) \tag{10.33}$$

where we are using the notation

$$\delta(z) \equiv \delta(x_1) \otimes \delta(x_2) \otimes \delta(p_1) \otimes \delta(p_2),$$
$$\delta'(z) \equiv \delta'(x_1) \otimes \delta'(x_2) \otimes \delta'(p_1) \otimes \delta'(p_2).$$

Expanding the function $\sin(px/2\hbar)$ in a Taylor series at the origin, we get

$$\chi_{\mathrm{BJ}}(z) = 1 + \sum_{k=1}^{\infty} \frac{(-1)^k}{(2k+1)!} \left(\frac{px}{2\hbar}\right)^{2k}$$

and hence, observing that $(px)^{2k}\delta'(z) = 0$ for $k > 1$,

$$a_\sigma(z)\chi_{\mathrm{BJ}}(z) = a_\sigma(z)\left(1 - \frac{(px)^2}{24\hbar^2}\right). \tag{10.34}$$

Comparing the expressions defining respectively the Weyl and Born–Jordan quantizations of a, it follows that the difference $\Delta(a) = \mathrm{Op_W}(a) - \mathrm{Op_{BJ}}(a)$ is given by

$$\begin{aligned}\Delta(a)\psi &= \left(\frac{1}{2\pi\hbar}\right)^2 \frac{1}{24\hbar^2} \int a_\sigma(z)(px)^2 \widehat{T}(z)\psi dz \\ &= \frac{\hbar^2}{12} \int \delta'(z)(px)^2 \widehat{T}(z)\psi dz.\end{aligned}$$

Using the elementary properties of the Dirac function we have

$$\delta'(z)(px)^2 = 2\delta(z) \tag{10.35}$$

and hence

$$\Delta(a)\psi = \frac{\hbar^2}{6} \int \delta(z)\widehat{T}(z)\psi d^4 z = \tfrac{1}{6}\hbar^2\psi$$

the second equality because

$$\delta(z)\widehat{T}(z) = \delta(z)e^{-\frac{i}{\hbar}\sigma(\widehat{z},z)} = \delta(z).$$

Since $\ell^2 = \ell_1^2 + \ell_2^2 + \ell_3^2$ a similar argument for the terms ℓ_1^2 and ℓ_2^2 shows that

$$(\widehat{\ell^2})_{\mathrm{W}} - (\widehat{\ell^2})_{\mathrm{BJ}} = \tfrac{1}{2}\hbar^2, \tag{10.36}$$

hence, taking (10.32) into account:

$$(\widehat{\ell^2})_{\mathrm{QM}} = (\widehat{\ell^2})_{\mathrm{W}} - \tfrac{3}{2}\hbar^2 = (\widehat{\ell^2})_{\mathrm{BJ}} - \hbar^2. \tag{10.37}$$

Summarizing these observations, we are in the following situation: as observed by Dahl and Springborg we have

$$\langle\psi|(\widehat{\ell^2})_{\mathrm{W}}|\psi\rangle = \tfrac{3}{2}\hbar^2 \tag{10.38}$$

while

$$\langle\psi|(\widehat{\ell^2})_{\mathrm{BJ}}|\psi\rangle = \hbar^2 \tag{10.39}$$

which is the result obtained by Dahl and Springborg [5] using an averaging procedure over what they call a "classical subspace". The Born–Jordan quantization procedure thus allows us to recover the value of the Bohr orbital angular momentum $\hbar^2$.

References

1. P. Boggiatto, G. De Donno, A. Oliaro, Time-frequency representations of Wigner type and pseudo-differential operators. Trans. Am. Math. Soc. **362**(9), 4955–4981 (2010)
2. P. Boggiatto, B.K. Cuong, G. De Donno, A. Oliaro, Weighted integrals of Wigner representations. J. Pseudo-Differ. Oper. Appl. (2010)
3. P. Boggiatto, G. Donno, A. Oliaro, Hudson's theorem for τ-Wigner transforms. Bull. London Math. Soc **45**(6), 1131–1147 (2013)
4. J.P. Dahl, W.P. Schleich, Concepts of radial and angular kinetic energies. Phys. Rev. A **65**(2), 022109 (2002)
5. J.P. Dahl, M. Springborg, Wigner's phase space function and atomic structure: I. The hydrogen atom ground state. Mol. Phys. **47**(5), 1001–1019 (1982)
6. M. de Gosson, Symplectic covariance properties for Shubin and Born–Jordan pseudo-differential operators. Trans. Am. Math. Soc. **365**, (2013)
7. H. Jeffreys, B. Jeffreys, *Methods of Mathematical Physics*, 3rd edn. (Cambridge University Press, Cambridge, 1972)
8. L. Pauling, *General Chemistry*, 3rd edn. (W.H. Freeman & Co., 1970), p. 125
9. J.R. Shewell, On the formation of quantum-mechanical operators. Am. J. Phys. **27**, 16–21 (1959)

Chapter 11
Weak Values and the Reconstruction Problem

The notions of weak measurements and values has become a very popular topic, with many fascinating ramifications (e.g. time-symmetric quantum mechanics, the theory of mutually unbiased bases, superoscillations, retrocausality, to name a few). It is also the subject of many ontological and philosophical debates, which are far from being resolved, and which we do not address here. In this chapter we will discuss the theory of weak values from a mathematical—and hence rigorous—point of view, and somewhat extend previous results of ours to the case of Born–Jordan quantization. We will in particular focus on the reconstruction problem, which is related to the Pauli conjecture about the determination of a state knowing its position and momentum. We will see that in the Born–Jordan case, this question leads to a—yet!—unsolved problem.

11.1 The Notion of Weak Value

The notion of weak value and weak measurement was introduced by Aharonov, Albert, Bergmann, Lebowitz and further developed by Vaidman; a non-exhaustive list of references is [1–8]. We begin by giving the definition of a weak values; we there after discuss some motivations.

11.1.1 Motivation and Definition

Let $\widehat{A}$ be a self-adjoint operator having eigenvalues $a_1, a_2, \dots$ with corresponding eigenfunctions $\psi_1, \psi_2, \dots$. In an ideal measurement (or "von Neumann measurement", as it is also called) the expectation value of $\widehat{A}$ in a quantum state $|\psi\rangle$ is given by the familiar expression

$$\langle\widehat{A}\rangle^{\psi} = \frac{\langle\psi|\widehat{A}|\psi\rangle}{\langle\psi|\psi\rangle}; \tag{11.1}$$

M.A. de Gosson, *Born–Jordan Quantization*, Fundamental Theories of Physics 182, DOI 10.1007/978-3-319-27902-2_11

if the sequence of eigenvalues lies in some interval $[a_{\min}, a_{\max}]$ then we will have $a_{\min} \leq \langle \widehat{A} \rangle^{\psi} \leq a_{\max}$. In fact, if one performs the ideal measurement the outcome will always be one of the eigenvalues λ_j, and the probability of this outcome is $|\lambda_j|^2/||\psi_j||^2$ where λ_j is the coefficient of ψ_j in the Fourier expansion $\psi = \sum_j \lambda_j \psi_j$. Moreover the system will be left in the state ψ_j after the measurement yielding the value a_j ("wavefunction collapse"). Assuming that the coupling with the measuring device is so small that the change of quantum state due to the interaction can be neglected. The eigenvalues are then not fully resolved, and the system is left in a superposition of these unresolved states. Now, if an appropriate post-selection is made, this superposition can interfere to produce a measurement result which can be significantly outside the range of the eigenvalues of the observable $\widehat{A}$. This is achieved by making the coupling with the measuring device so small that the change of quantum state due to the interaction can be neglected. The post-selection can then be accomplished by making an ideal measurement of some other observable $\widehat{B}$ and selecting one particular outcome. Thus, the post-selected state $|\phi\rangle$ is an eigenstate of $\widehat{B}$ which can be expressed as a linear combination of the eigenstates of $\widehat{A}$.

Definition 1 Let ϕ and ψ be square integrable functions. If $\langle \phi | \psi \rangle \neq 0$ the *weak value* of $\widehat{A}$ with respect to the states $|\phi\rangle$ and $|\psi\rangle$ is the complex number

$$\langle \widehat{A} \rangle_{\text{weak}}^{\phi,\psi} = \frac{\langle \phi | \widehat{A} | \psi \rangle}{\langle \phi | \psi \rangle} = \frac{\int \phi^*(x) \widehat{A} \psi(x) d^n x}{\int \phi^*(x) \psi(x) d^n x}. \tag{11.2}$$

Observe that the complex number $\langle \widehat{A} \rangle_{\text{weak}}^{\phi,\psi}$ really only depends on the states $|\phi\rangle$ and $|\psi\rangle$ since, by sequilinearity, we have $\langle \widehat{A} \rangle_{\text{weak}}^{\lambda\phi,\lambda\psi} = \langle \widehat{A} \rangle_{\text{weak}}^{\phi,\psi}$ for every complex number $\lambda \neq 0$. Also notice that we are loosely talking about the "quantum observable $\widehat{A}$" without specifying for the moment to which classical observable it is associated with; the definition above is thus quite general.

Here is a suggestive description of the relation between the notion of weak value and *retrocausality*. Assume that at a time t_{in} an observable $\widehat{A}$ is measured and a non-degenerate eigenvalue was found: $|\psi(t_{\text{in}})\rangle = |\widehat{A} = a\rangle$; similarly at a later time t_{fin} a measurement of another observable $\widehat{B}$ yields $|\phi(t_{\text{fin}})\rangle = |\widehat{B} = b\rangle$. Let t be some intermediate time: $t_{\text{in}} < t < t_{\text{fin}}$. Following the time-symmetric approach to quantum mechanics (see the review in [4]), at this intermediate time the system is described by the *two* wavefunctions

$$\psi(t) = U(t, t_{\text{in}})\psi(t_{\text{in}}),\ \phi(t) = U(t, t_{\text{fin}})\phi(t_{\text{fin}}) \tag{11.3}$$

where $U(t, t')$ is the unitary operator defined by the Schrödinger equation

$$i\hbar \frac{d}{dt} U(t, t') = \widehat{H} U(t, t'),\ U(t, t) = I_{\text{d}} \tag{11.4}$$

and $\widehat{H}$ is the quantum Hamiltonian (assumed to be time-independent); that is $U(t,t') = e^{-i\widehat{H}(t-t')/\hbar}$. Notice that the wavefunction $\phi(t)$ travels *backwards* in time since $t < t_{\text{fin}}$. Consider now the superposition of the two states $|\psi(t)\rangle$ and $|\phi(t)\rangle$; obviously the expectation value $\langle\widehat{A}\rangle^{\psi(t)+\phi(t)}$ of this superposition is given by (dropping the reference to t)

$$\begin{aligned}\langle\widehat{A}\rangle^{\psi+\phi} &= \frac{||\phi||}{||\phi+\psi||}\langle\widehat{A}\rangle^{\phi} + \frac{||\psi||}{||\phi+\psi||}\langle\widehat{A}\rangle^{\psi} \\ &+ \frac{2}{||\phi+\psi||}\text{Re}\left(\langle\phi|\psi\rangle\langle\widehat{A}\rangle^{\psi,\phi}_{\text{weak}}\right)\end{aligned}$$

that is

$$\begin{aligned}\langle\widehat{A}\rangle^{\psi+\phi} &= \frac{||\phi||}{||\phi+\psi||}\langle\widehat{A}\rangle^{\phi} + \frac{||\psi||}{||\phi+\psi||}\langle\widehat{A}\rangle^{\psi} \\ &+ \frac{2}{||\phi+\psi||}\text{Re}\langle\phi|\widehat{A}|\psi\rangle.\end{aligned}$$

This makes apparent the importance of the cross-term $\langle\phi(t)|\widehat{A}|\psi(t)\rangle$ appearing in the definition (11.2) of weak values. This observation is the starting point of time-symmetric quantum mechanics, where the actual quantum state consists of *two* wavefunctions $\phi(t)$ and $\psi(t)$ (see Aharonov and Vaidman [6, 8], Aharonov et al. [4]).

11.1.2 Weak Values and the Cohen Class: The Phase Space Approach

Another way of making explicit the relation between interference effects and weak values is to use the Cohen class formalism. Returning to the situation described above, and dropping again for notational simplicity the reference to the time t, the Wigner distribution of the state $|\psi\rangle + |\phi\rangle$ is

$$Q(\psi+\phi) = Q\psi + Q\phi + 2\text{Re}Q(\psi,\phi) \tag{11.5}$$

where

$$Q(\psi,\phi) = \text{Wig}(\psi,\phi) * \theta \tag{11.6}$$

for some $\theta \in \mathcal{S}'(\mathbb{R}^{2n})$ (see Chap. 7 for the definition of the Cohen class). It follows that we can write

$$\text{Re}\left(\langle\phi|\psi\rangle\langle\widehat{A}\rangle^{\psi,\phi}_{\text{weak}}\right) = \int \text{Re}(Q(\phi,\psi)(z)a(z))d^{2n}z$$

where a is the symbol of the operator $\widehat{A}$ in the quantization corresponding to Q (see Proposition 17 in Chap. 7). These relations make obvious the fact that the weak value is due to the interference term coming from the quasi-distribution Q via the classical observable a such that $\widehat{A} = \mathrm{Op}_Q(a)$.

Proposition 2 *Let Q be an element of the Cohen class such that $Q : \mathcal{S}(\mathbb{R}^n) \times \mathcal{S}(\mathbb{R}^n) \longrightarrow \mathcal{S}(\mathbb{R}^{2n})$. Assume that $\widehat{A} = \mathrm{Op}_Q(a)$ for some $a \in \mathcal{S}'(\mathbb{R}^{2n})$ and that $\widehat{A} : \mathcal{S}(\mathbb{R}^n) \longrightarrow \mathcal{S}(\mathbb{R}^n)$. We have, for all $(\psi, \phi) \in L^2(\mathbb{R}^n) \times L^2(\mathbb{R}^n)$:*

$$\langle \widehat{A} \rangle_{\text{weak}}^{\phi,\psi} = \frac{1}{\langle \phi | \psi \rangle} \int a(z) Q(\psi, \phi)(z) d^{2n} z. \tag{11.7}$$

Proof In view of Proposition 17 in Chap. 7 the operator $\widehat{A}$ is related to Q by the formula (7.33)

$$\langle \widehat{A}\psi, \phi^* \rangle = \langle\langle a, Q(\psi, \phi) \rangle\rangle$$

that is

$$\int \widehat{A}\psi(x) \phi^*(x) d^n x = \int a(z) Q(\psi, \phi)(z) d^{2n} z$$

hence formula (11.7) since $\phi, \psi \in L^2(\mathbb{R}^n)$. ∎

11.2 A Complex Probability Density

Weak values can be interpreted as the average of the *classical* observable with respect to a complex quasi-probability distribution; this approach was initiated in [12, 13].

11.2.1 A General Result Using the Cohen Class

We are going to see that the cross-term $Q(\psi, \phi)$ has a simple probabilistic interpretation. We assume that Q satisfies the marginal properties

$$\int Q(\phi, \psi)(z) d^n p = \psi^*(x) \phi(x) \tag{11.8}$$

$$\int Q(\phi, \psi)(z) d^n x = \widehat{\psi}^*(p) \widehat{\phi}(p). \tag{11.9}$$

This is the case if and only if the symplectic Fourier transform of the Cohen kernel θ satisfies the two conditions:

$$\theta_\sigma(0,p) = \theta_\sigma(x,0) = (2\pi\hbar)^{-n}. \tag{11.10}$$

(Proposition 14 in Chap. 7).

Proposition 3 *Assume that $\langle\psi|\phi\rangle \neq 0$. (i) The function $\rho_Q^{\phi,\psi}$ on $\mathbb{R}^{2n}$ defined by*

$$\rho_Q^{\phi,\psi}(z) = \frac{Q(\psi,\phi)(z)}{\langle\psi|\phi\rangle} \tag{11.11}$$

is a complex probability distribution:

$$\int \rho_Q^{\phi,\psi}(z)d^{2n}z = 1 \tag{11.12}$$

and the corresponding (complex) marginals are given by

$$\int \rho_Q^{\phi,\psi}(z)d^n p = \frac{\phi^*(x)\psi(x)}{\langle\phi|\psi\rangle} \tag{11.13}$$

$$\int \rho_Q^{\phi,\psi}(z)d^n x = \frac{\widehat{\phi}(p)^*\widehat{\psi}(p)}{\langle\phi|\psi\rangle}. \tag{11.14}$$

Proof Since Q satisfies the marginal conditions we have

$$\int \rho_Q^{\phi,\psi}(z)d^n p = \frac{1}{\langle\phi|\psi\rangle}\int Q(\psi,\phi)(z)d^n p = \frac{\phi^*(x)\psi(x)}{\langle\phi|\psi\rangle}$$

and similarly

$$\int \rho_Q^{\phi,\psi}(z)d^n x = \frac{1}{\langle\phi|\psi\rangle}\int Q(\psi,\phi)(z)d^n x = \frac{\widehat{\phi}(p)^*\widehat{\psi}(p)}{\langle\phi|\psi\rangle}.$$

It follows that

$$\int \rho_Q^{\phi,\psi}(z)d^{2n}z = \frac{1}{\langle\phi|\psi\rangle}\int \phi^*(x)\psi(x)d^n x = 1$$

hence $\rho_Q^{\phi,\psi}$ is indeed a (complex) probability density with marginals (11.13) and (11.14). ∎

We can restate (11.7) in terms of the complex quasi-distribution $\rho_Q^{\phi,\psi}$: let $\widehat{A} = \operatorname{Op}_Q(a)$; the weak value (11.7) is then given by the formula:

$$\langle \widehat{A} \rangle^{\phi,\psi}_{\text{weak}} = \int \rho^{\phi,\psi}_{Q}(z)a(z)d^{2n}z \tag{11.15}$$

which shows that in a sense the weak value $\langle \widehat{A} \rangle^{\phi,\psi}_{\text{weak}}$ is the symbol a averaged by $\rho^{\phi,\psi}_{Q}$.

Observe that the real and imaginary parts of the complex quasi-distribution $\rho^{\phi,\psi}_{Q}$ satisfy the relations

$$\int \text{Re}\rho^{\phi,\psi}_{Q}(z)d^{2n}z = 1, \int \text{Im}\rho^{\phi,\psi}_{Q}(z)d^{2n}z = 0.$$

The physical meaning of these formulas is the following (Aharonov and Vaidman [7], Chap. 14): the readings of the pointer of the measuring device will cluster around

$$\text{Re}\langle \widehat{A}_Q \rangle^{\phi,\psi}_{\text{weak}} = \int \text{Re}(a(z)\rho^{\phi,\psi}_{Q}(z))d^{2n}z$$

while

$$\text{Im}\langle \widehat{A}_Q \rangle^{\phi,\psi}_{\text{weak}} = \int \text{Im}(a(z)\rho^{\phi,\psi}_{Q}(z))d^{2n}z$$

measures the average shift in the variable conjugate to the pointer variable.

Example 4 Assume that the Cohen kernel $\theta = \delta$ (the Dirac distribution on $\mathbb{R}^{2n}$). Then $Q(\psi, \phi) = \text{Wig}(\psi, \phi)$ is the usual cross-Wigner distribution, and we have

$$\rho^{\phi,\psi}_{\text{Wig}}(z) = \frac{\text{Wig}(\psi, \phi)}{\langle \psi | \phi \rangle} \tag{11.16}$$

so that

$$\langle \widehat{A}_{\text{W}} \rangle^{\phi,\psi}_{\text{weak}} = \int a(z)\rho^{\phi,\psi}_{\text{Wig}}(z)d^{2n}z. \tag{11.17}$$

(see de Gosson and de Gosson [13]).

11.2.2 The Born–Jordan Case

We now specialize our discussion to the case

$$Q(\psi, \phi) = \text{Wig}_{\text{BJ}}(\psi, \phi) = \text{Wig}(\psi, \phi) * \theta_{\text{BJ}} \tag{11.18}$$

where θ_{BJ} is the Cohen kernel defined by

$$\theta_{\mathrm{BJ}} = (2\pi\hbar)^{-n} F_\sigma \chi_{\mathrm{BJ}}$$
$$\chi_{\mathrm{BJ}}(z) = \operatorname{sinc}(px/2\pi\hbar).$$

(Definitions 1 and 2 in Chap. 8). Recalling that $\mathrm{Wig}_{\mathrm{BJ}}(\psi, \phi)$ satisfies the marginal conditions (see formula (8.7)) we may apply the discussion above (see [14]). In this case, the complex quasi-probability density (11.11) becomes

$$\rho_{\mathrm{BJ}}^{\phi,\psi}(z) = \frac{\mathrm{Wig}_{\mathrm{BJ}}(z)}{\langle\phi|\psi\rangle}; \tag{11.19}$$

when $\widehat{A}_{\mathrm{BJ}} = \mathrm{Op}_{\mathrm{BJ}}(a)$ this leads to the expression

$$\langle\widehat{A}_{\mathrm{BJ}}\rangle_{\mathrm{weak}}^{\phi,\psi} = \int a(z)\rho_{\mathrm{BJ}}^{\phi,\psi}(z)d^{2n}z. \tag{11.20}$$

Notice that for a given operator it matters which quantization scheme one uses. While the Born–Jordan and Weyl quantizations are the same for the position and momentum variables x_j, p_k, we get different results for $\langle\widehat{A}_{\mathrm{BJ}}\rangle_{\mathrm{weak}}^{\phi,\psi}$ and $\langle\widehat{A}_{\mathrm{W}}\rangle_{\mathrm{weak}}^{\phi,\psi}$ as soon as one consider more complicated cases. Let us illustrate this on the example of the squared angular momentum already considered in Sect. 10.3.3 of Chap. 10.

Example 5 Consider $\ell_3 = x_1p_2 - x_2p_1$. The Born–Jordan and Weyl quantizations of the square ℓ_3^2 are related by

$$(\widehat{\ell^2})_{\mathrm{W}} - (\widehat{\ell^2})_{\mathrm{BJ}} = \tfrac{1}{2}\hbar^2. \tag{11.21}$$

It follows that the weak values $\langle(\widehat{\ell^2})_{\mathrm{BJ}}\rangle_{\mathrm{weak}}^{\phi,\psi}$ and $\langle(\widehat{\ell^2})_{\mathrm{W}}\rangle_{\mathrm{weak}}^{\phi,\psi}$ satisfy

$$\langle(\widehat{\ell^2})_{\mathrm{W}}\rangle_{\mathrm{weak}}^{\phi,\psi} - \langle(\widehat{\ell^2})_{\mathrm{BJ}}\rangle_{\mathrm{weak}}^{\phi,\psi} = \tfrac{1}{2}\hbar^2.$$

11.3 The Reconstruction Problem

In 1958 Wolfgang Pauli conjectured that one can reconstruct a quantum state $|\psi\rangle$ knowing its position and momentum; this conjecture was later disproved; see the discussion in Corbett [10]. In fact, even the knowledge of the position and momentum probability densities $|\psi(x)|^2$ and $|\widehat{\psi}(p)|^2$ is not sufficient (see the discussion in Lundeen et al. [16]). Mathematically this means that the knowledge of the marginal of the quasi-distribution $W\psi$ is not enough to determine $W\psi$. It turns out that, however, the knowledge of the cross-Wigner distribution $W(\psi, \phi)$ and of one of the two functions ψ or ϕ unambiguously determines the other. The relation of this property with the notion of weak value is obvious in view of our previous discussion of the complex quasi-density $\rho_{\mathrm{W}}^{\phi,\psi}$. In this section we begin by presenting a general procedure for the reconstruction problem for elements of the Cohen class satisfying the Moyal identity. This method does not, however, apply to the Born–Jordan case; the latter will be considered separately. We will see that it leads to an open problem.

11.3.1 A General Reconstruction Formula

The inversion formula is well-known when Q is the cross-Wigner transform [11, 15]; see formula (7.18) in Chap. 7. We prove here a more general result valid for all members of the Cohen class which satisfy the Moyal identity. In view of Proposition 16 in Chap. 7 the generalized Moyal identity

$$\langle\langle Q(\psi,\phi)|Q(\psi',\phi')\rangle\rangle = \left(\tfrac{1}{2\pi\hbar}\right)^n \langle\psi|\psi'\rangle\langle\phi|\phi'\rangle^* \tag{11.22}$$

holds if and only if the symplectic Fourier transform of the Cohen kernel of Q satisfies

$$|\theta_\sigma(z)| = (2\pi\hbar)^{-n} \tag{11.23}$$

(in which case Q automatically satisfies the marginal conditions (11.8)–(11.9) since the conditions (11.10) are then verified).

Proposition 6 *Assume that the element Q of the Cohen class satisfies the Moyal identity* (11.22). *Let $\phi,\gamma\in L^2(\mathbb{R}^n)$ be two non-orthogonal states: $\langle\phi|\gamma\rangle\neq 0$. Then*

$$\psi(x) = \frac{1}{\langle\gamma|\phi\rangle}\int Q_\sigma(\psi,\phi)(z_0)\widehat{T}_Q(z_0)\gamma(x)d^{2n}z_0. \tag{11.24}$$

Proof Let us set

$$\chi(x) = \int Q_\sigma(\psi,\phi)(z_0)\widehat{T}_Q(z_0)\gamma(x)d^{2n}z_0$$

and choose an arbitrary $\alpha\in\mathcal{S}(\mathbb{R}^n)$. We have

$$\begin{aligned}\langle\alpha|\chi\rangle &= \int Q_\sigma(\psi,\phi)(z_0)\langle\alpha|\widehat{T}_Q(z_0)\gamma\rangle d^{2n}z_0\\ &= \int Q_\sigma(\psi,\phi)(z_0)\langle\widehat{T}_Q(z_0)\gamma|\alpha\rangle^* d^{2n}z_0\\ &= (2\pi\hbar)^n\int Q_\sigma(\psi,\phi)(z_0)Q_\sigma(\alpha,\gamma)^*(z_0)d^{2n}z_0\\ &= (2\pi\hbar)^n\langle\langle Q_\sigma(\alpha,\gamma),Q_\sigma(\psi,\phi)\rangle\rangle.\end{aligned}$$

Applying Moyal's formula to the last equality we get $\langle\alpha|\chi\rangle = \langle\alpha|\psi\rangle\langle\gamma|\phi\rangle$; since this identity holds for all $\alpha \in \mathcal{S}(\mathbb{R}^n)$ we have $\chi = \langle\gamma|\phi\rangle\psi$ almost everywhere, which proves formula hence (11.24). ∎

Notice that if we choose in particular $\gamma = \phi$ then we get

$$\psi(x) = \frac{1}{||\phi||^2}\int Q_\sigma(\psi,\phi)(z_0)\widehat{T}_Q(z_0)\phi(x)d^{2n}z_0. \tag{11.25}$$

In particular, if $Q(\psi,\phi)$ is the cross-Wigner transform, then we get

$$\psi(x) = \frac{1}{\langle\gamma|\phi\rangle}\int \mathrm{Amb}(\psi,\phi)(z_0)\widehat{T}(z_0)\gamma(x)d^{2n}z_0. \tag{11.26}$$

Using a symplectic Fourier transform together with formula (6.26) in Chap. 6, we recover the formula (7.18)

$$\psi(x) = \frac{2^n}{\langle\phi|\gamma\rangle}\int \mathrm{Wig}(\psi,\phi)(z_0)\widehat{T}_{\mathrm{GR}}(z_0)\gamma(x)d^{2n}z_0.$$

which was proven in Chap. 7.

Example 7 For $\tau \neq \frac{1}{2}$ the Cohen kernel

$$\theta_{(\tau)}(z) = \frac{2^n}{|2\tau-1|^n}e^{\frac{2i}{\hbar(2\tau-1)}px} \tag{11.27}$$

has Fourier transform

$$\widehat{\theta}_{(\tau)}(z) = \left(\tfrac{1}{2\pi\hbar}\right)^n \frac{(2\tau-1)^n}{|2\tau-1|^n}e^{-\frac{i(2\tau-1)}{2\hbar}px}.$$

and hence $|\widehat{\theta}_{(\tau)}(z)| = (2\pi\hbar)^{-n}$. It follows that the τ-cross-Wigner distribution $W_\tau(\psi,\phi)$ satisfies the Moyal identity.

11.3.2 The Born–Jordan Case: An Unsolved Problem

The reconstruction problem has unexpected difficulties in the Born–Jordan case. We would like here to reconstruct ψ from the knowledge of the Born–Jordan cross-transform $\mathrm{Wig}_{\mathrm{BJ}}(\psi,\phi)$, knowing ϕ. However, as already noticed in Chap. 8 (Sect. 8.1.2) the Moyal identity does not hold for the quasi-distribution $\mathrm{Wig}_{\mathrm{BJ}}(\psi,\phi)$; it is therefore not possible to apply Proposition 6, whose proof precisely relies on the validity of the Moyal identity. We are thus led here to a problem which is, at the time of writing, *unsolved.* This difficulty clearly shows that Born–Jordan quantization has very peculiar features, which might be related to the fact that "real" quantum mechanics isn't, aftre all, that simple (as opposed to the conventional Weyl approach). It might very well be that the mathematical difficulty is related to a division problem, as was the case when we studied the inveribility of the Born–Jordan correspondence in Chap. 8 (Sect. 8.3). Recall that $\mathrm{Wig}_{\mathrm{BJ}}(\psi, \phi)$ and $\mathrm{Wig}(\psi, \phi)$ are related by a convolution formula, namely

$$\mathrm{Wig}_{\mathrm{BJ}}(\psi, \phi) = \mathrm{Wig}(\psi, \phi) * \theta_{\mathrm{BJ}}$$

where θ_{BJ} is the Born–Jordan kernel. If we were able to "invert" this relation, and prove that the usual cross-Wigner transform $\mathrm{Wig}(\psi, \phi)$ can be expressed in some way in terms of $\mathrm{Wig}_{\mathrm{BJ}}(\psi, \phi)$, we would be led back to the standard reconstruction problem. But, for the moment being, this is an open problem.

References

1. Y. Aharonov, P.G. Bergmann, J. Lebowitz, Time symmetry in the quantum process of measurement. Phys. Rev. B **134**, B1410–B1416 (1964)
2. Y. Aharonov, D.Z. Albert, L. Vaidman, How the result of a measurement of a component of the spin of a spin-$\frac{1}{2}$ particle can turn out to be 100. Phys. Rev. Lett. **60**(14), 1351–1354 (1988)
3. Y. Aharonov, A. Botero, Quantum averages of weak values. Phys. Rev. A **72**, 052111 (2005)
4. Y. Aharonov, S. Popescu, J. Tollaksen, A time-symmetric formulation of quantum mechanics. Phys. Today 27–32 (2010)
5. Y. Aharonov, L. Vaidman, Properties of a quantum system during the time interval between two measurements. Phys. Rev. A **41**(1), 11–20 (1990)
6. Y. Aharonov, L. Vaidman, Complete description of a quantum system at a given time. J. Phys. A: Math. Gen. **24**, 2315–2328 (1991)
7. Y. Aharonov, L. Vaidman, The two-state vector formalism: an updated review. Lect. Notes. Phys. **734**, 399–447 (2008)
8. Y. Aharonov, L. Vaidman, The two-state vector formalism: an updated review. Lect. Notes Phys. **734**, 399–447 (2008). [quant-ph/0105101v2]
9. M.V. Berry, P. Shukla, Typical weak and superweak values. J. Phys. A: Math. Theor. **43**, 354024 (2010)
10. J.V. Corbett, The Pauli problem, state reconstruction and quantum-real numbers. Rep. Math. Phys. **57**(1), 53–68 (2006)
11. M. de Gosson, *Symplectic Methods in Harmonic Analysis and in Mathematical Physics* (Birkhäuser, 2011)
12. M. de Gosson, S. de Gosson, The reconstruction problem and weak quantum values. J. Phys. A: Math. Theor. **45**(11), 115305 (2012)
13. M. de Gosson, S. de Gosson, Weak values of a quantum observable and the cross-Wigner distribution. Phys. Lett. A **376**(4), 293–296 (2012)
14. M. de Gosson, L.D. Abreu, Weak values and Born-Jordan quantization. Quantum Theory: Reconsiderations Found. **6**(1508), 156–161 (2012)
15. K. Gröchenig, *Foundations of Time-Frequency Analysis* (Birkhäuser, Boston, 2000)
16. J.S. Lundeen, B. Sutherland, A. Patel, C. Stewart, C. Bamber, Direct measurement of the quantum wavefunction. Nature **474**(7350), 188–191 (2011)
17. W. Pauli, *Die allgemeinen Prinzipen der Wellenmechanik, Handbuch der Physik*, vol. 5 (Springer, Berlin, 1958)
18. N.W.M. Ritchie, J.G. Story, R.G. Hulet, Realization of a measurement of a "weak value". Phys. Rev. Lett. **66**(9), 1107–1110 (1991)

Part III
Some Advanced Topics

Chapter 12
Metaplectic Operators

The metaplectic group is a unitary representation of the double cover of the symplectic group; it is thus characterized by the exactness of the sequence

$$0 \longrightarrow \mathbb{Z}_2 \longrightarrow \mathrm{Mp}(n) \longrightarrow \mathrm{Sp}(n) \longrightarrow 0.$$

The interest of the metaplectic group comes from the fact that it is a maximal group of symmetries for Weyl operators ("symplectic covariance"), and contains subgroups under which Shubin and Born–Jordan operators are symplectically covariant.

The metaplectic representation of the symplectic group has a rather long history, and is a subject of interest both for mathematicians and physicists. The germ of the idea of the metaplectic representation is found in van Hove [16]. It then appears in the work of Segal [14], Shale [15], and Weil [18]. The theory has been subsequently developed by Buslaev [1], Maslov [12], Leray [11], Reiter [13], and the author [3, 4, 6, 10]. For somewhat different presentations of the theory see Folland [2] and Wallach [17].

12.1 The Symplectic Group Sp(*n*)

For an extensive study of the symplectic group with complete proofs we refer to de Gosson [6]. We recall that the standard symplectic form on the phase space $\mathbb{R}^{2n}$ is defined by

$$\sigma(z, z') = \sum_{j=1}^{n} p_j x'_j - p'_j x_j = Jz \cdot z'$$

where $J = \begin{pmatrix} 0_{n\times n} & I_{n\times n} \\ -I_{n\times n} & 0v \end{pmatrix}$.

M.A. de Gosson, *Born–Jordan Quantization*, Fundamental Theories of Physics 182, DOI 10.1007/978-3-319-27902-2_12

Definition 1 The symplectic group $\mathrm{Sp}(n)$ is the group of all linear automorphisms S of $\mathbb{R}^{2n}$ consisting of all linear automorphisms S such that $\sigma(Sz, Sz') = \sigma(z, z')$ for all $(z, z') \in \mathbb{R}^{2n} \times \mathbb{R}^{2n}$; equivalently $S^T JS = SJS^T = J$.

The group $\mathrm{Sp}(n)$ is a connected Lie group, invariant under transposition; it is generated by the subset consisting of all free symplectic matrices. Recall from Chap. 4 that a $2n \times 2n$ real matrix S is said to be free if we can write it in block-matrix form

$$S = \begin{pmatrix} A & B \\ C & D \end{pmatrix} \text{ with } \det B \neq 0. \tag{12.1}$$

Since the inverse of S is

$$S^{-1} = \begin{pmatrix} D^T & -B^T \\ -C^T & A^T \end{pmatrix} \tag{12.2}$$

it follows that S is free if and only if S^{-1} is free. The notion of free symplectic matrix allows a quite interesting description of a class of generators of the group $\mathrm{Sp}(n)$, using the following result:

Proposition 2 *Every $S \in \mathrm{Sp}(n)$ can be written as a product $S = S_1 S_2$ of two free symplectic matrices S_1 and S_2.*

The proof of this result is unexpectedly difficult (except in the case $n = 1$, where it is elementary); we refer to de Gosson [4, 6]. Defining the elementary symplectic matrices

$$V_{-P} = \begin{pmatrix} I & 0 \\ P & I \end{pmatrix}, \; M_L = \begin{pmatrix} L^{-1} & 0 \\ 0 & L^T \end{pmatrix} \tag{12.3}$$

with P symmetric and L invertible, an easy calculation shows that every free symplectic matrix (12.1) can be factorized as

$$S = V_{-DB^{-1}} M_{B^{-1}} J V_{-B^{-1}A}. \tag{12.4}$$

(In fact, the conditions $S^T JS = SJS^T = J$ compel both $P = DB^{-1}$ and $Q = B^{-1}A$ to be symmetric, as a straightforward calculation of block-matrices show.) It follows that:

Proposition 3 *(i) The set of real $2n \times 2n$ matrices*

$$\{V_P, M_L, J : P = P^T, \det L \neq 0\}$$

generates the symplectic group $\mathrm{Sp}(n)$. *(ii) Symplectic matrices have determinant one.*

Proof (i) In view of Proposition 2 and the factorization (12.4) every $S \in \mathrm{Sp}(n)$ can be written $S = S_1 S_2$ where

$$S_1 = V_{-D_1 B_1^{-1}} M_{B_1^{-1}} J V_{-B_1^{-1} A_1}$$
$$S_2 = V_{-D_2 B_2^{-1}} M_{B_2^{-1}} J V_{-B_2^{-1} A_2}$$

hence S is a product of symplectic matrices of the type V_P, M_L, J. (ii) We have $\det S = \det S_1 \det S_2$ and $\det S_1 = \det S_2 = 1$ since the determinants of the matrices V_P, M_L, J obviously are equal to one. ∎

Let us now pursue our discussion of the notion of generating function initiated in Chap. 4.

Definition 4 The generating function of the free symplectic matrix (12.1) is the quadratic form

$$W(x, x') = \tfrac{1}{2}DB^{-1}x^2 - B^{-1}x \cdot x' + \tfrac{1}{2}B^{-1}Ax'^2. \tag{12.5}$$

When the generating function is given, we will write $S = S_W$.

W is called a generating function, because its datum completely defines S:

$$(x, p) = S(x', p') \Longleftrightarrow \begin{cases} p = \nabla_x W(x, x') \\ p' = -\nabla_{x'} W(x, x') \end{cases}; \tag{12.6}$$

in fact the relations $p = \nabla_x W(x, x')$ and $p' = -\nabla_{x'} W(x, x')$ are equivalent to

$$p = DB^{-1}x - (B^{-1})^T x', \; p' = B^{-1}x - B^{-1}Ax';$$

solving these equations in x, p indeed yields

$$x = Ax' + Bp', \; p = Cx' + Dp'.$$

The generating functions of S_W and of its inverse S_W^{-1} are given by:

$$S_W^{-1} = S_{W^*}, \; W^*(x, x') = -W(x', x)$$

as is easily checked using (12.2).

We observe that, conversely, every real quadratic form

$$W(x, x') = \tfrac{1}{2}Px^2 - Lx \cdot x' + \tfrac{1}{2}Qx'^2 \tag{12.7}$$

with $P = P^T$, $Q = Q^T$, and $\det L \neq 0$, generates a free symplectic matrix, namely:

$$S_W = \begin{pmatrix} L^{-1}Q & L^{-1} \\ PL^{-1}Q - L^T & PL^{-1} \end{pmatrix} \tag{12.8}$$

(one can use the property $S_W^T J S_W = S_W J S_W^T = J$ to prove that $S_W \in \mathrm{Sp}(n)$). For instance, if $W(x, x') = x \cdot x'$ we have $S_W = J$.

12.1.1 The Metaplectic Representation of Sp(n)

The symplectic group is contractible to the maximal compact subgroup $U(n)$, image of the unitary group $U(n, \mathbb{C})$ by the monomorphism

$$A + iB \longmapsto \begin{pmatrix} A & -B \\ B & A \end{pmatrix}.$$

It follows that the fundamental group $\pi_1[\mathrm{Sp}(n)]$ is isomorphic to the integer group $(\mathbb{Z}, +)$ hence $\mathrm{Sp}(n)$ has covering groups $\mathrm{Sp}_q(n)$ of all orders $q = 1, 2, ..., \infty$. Among all these covering groups the double cover $\mathrm{Sp}_2(n)$ plays a very privileged role, because it can be (faithfully) represented by a group $\mathrm{Mp}(n)$ of unitary operators acting on $L^2(\mathbb{R}^n)$.

Definition 5 The group $\mathrm{Mp}(n) \equiv \mathrm{Sp}_2(n)$ is called the metaplectic group. We denote the natural covering projection by

$$\pi : \mathrm{Mp}(n) \longrightarrow \mathrm{Sp}(n).$$

The generic element of $\mathrm{Mp}(n)$ is denoted by $\widehat{S}$ and we write $S = \pi(\widehat{S})$.

The generators of $\mathrm{Mp}(n)$ can be described in many way. The simplest (but not necessarily the most tractable) is the following:

Proposition 6 *The group* $\mathrm{Mp}(n)$ *is generated by* $\widehat{J} = i^{-n/2}F$ *(*F *the* $\hbar$*-dependent Fourier transform on* $L^2(\mathbb{R}^n)$*), and the operators* $\widehat{V}_{-P}$ *and* $\widehat{M}_{L,m}$ *defined by*

$$\widehat{V}_{-P}\psi(x) = e^{\frac{i}{2\hbar}Px^2}\psi(x) \tag{12.9}$$

$$\widehat{M}_{L,m}\psi(x) = i^m\sqrt{|\det L|}\psi(Lx); \tag{12.10}$$

where the integer m *is chosen so that*

$$m\pi \equiv \arg\det L \mod 2\pi.$$

The projections of these operators on $\mathrm{Sp}(n)$ *are given by*

$$\pi(\widehat{J}) = J,\ \pi(\widehat{V}_{-P}) = V_{-P},\ \pi(\widehat{M}_{L,m}) = M_{L,m}.$$

(see de Gosson [6]). Another way is to define the analogue of free symplectic matrices in the metaplectic case (Leray [11], de Gosson [4, 6, 10]):

Proposition 7 *(i) The metaplectic group* $\mathrm{Mp}(n)$ *is generated by the Fourier integral operators* $\widehat{S}_{W,m} : \mathcal{S}(\mathbb{R}^n) \longrightarrow \mathcal{S}(\mathbb{R}^n)$ *associated to a generating function* (12.7) *by*

$$\widehat{S}_{W,m}\psi(x) = i^m\left(\frac{1}{2\pi i\hbar}\right)^{n/2}\sqrt{|\det L|}\int e^{\frac{i}{\hbar}W(x,x')}\psi(x')d^nx' \tag{12.11}$$

where the integer m is defined by

$$m\pi \equiv \arg \det L \mod 2\pi. \tag{12.12}$$

(ii) The operator $\widehat{S}_{W,m}$ *is invertible with inverse*

$$\widehat{S}_{W,m}^{-1} = \widehat{S}_{W^*,m^*},\ W^*(x,x') = -W(x',x),\ m^* = n - m. \tag{12.13}$$

(iii) The projection $\pi(\widehat{S}_{W,m})$ *is the free symplectic matrix* S_W *generated by* W.

As a consequence of the obvious relations

$$\widehat{V}_{-DB^{-1}}\psi(x) = e^{\frac{i}{2\hbar}DB^{-1}x^2}\psi(x) \tag{12.14}$$

$$\widehat{M}_{B^{-1},m}\psi(x) = i^m\sqrt{|\det B^{-1}|}\psi(B^{-1}x); \tag{12.15}$$

the operator $\widehat{S}_{W,m}$ can be factorized as a product

$$\widehat{S}_{W,m} = \widehat{V}_{-DB^{-1}}\widehat{M}_{B^{-1},m}\widehat{J}\widehat{V}_{-B^{-1}A} \tag{12.16}$$

(cf. the corresponding formula (12.4) for free symplectic matrices).

12.2 The Weyl Representation of Metaplectic Operators

Metaplectic operators are Weyl operators in their own right. However, the determination of the Weyl symbol of such an operator is rather technical.

12.2.1 The Symplectic Cayley Transform

We will use the following notation:

$$\mathrm{Sp}_0(n) = \{S \in \mathrm{Sp}(n) : \det(S - I) \neq 0\}$$

$$\mathrm{Sym}_0(n) = \{M \in \mathrm{Sym}(2n,\mathbb{R}) : \det(M - \tfrac{1}{2}J) \neq 0\}.$$

Let $S \in \mathrm{Sp}_0(n)$; by definition the symplectic Cayley transform (introduced in de Gosson [6–10]) of S is the symmetric matrix given by

$$M(S) = \tfrac{1}{2}J(S + I)(S - I)^{-1}. \tag{12.17}$$

The symmetry of $M(S)$ is readily verified using the relation $S^TJS = SJS^T = J$, which is equivalent to $S \in \mathrm{Sp}(n)$. It is clear that the mapping $M(\cdot)$ is a bijection

$\mathrm{Sp}_0(n) \longrightarrow \mathrm{Sym}_0(2n, \mathbb{R})$ and that the inverse of that bijection is given by

$$S = (M - \tfrac{1}{2}J)^{-1}(M + \tfrac{1}{2}J). \tag{12.18}$$

We have the properties

$$M(S^{-1}) = -M(S) \tag{12.19}$$

and, when in addition both S' and SS' are in $\mathrm{Sp}_0(n)$:

$$M(SS') = M(S) + (S^T - I)^{-1}J(M(S) + M(S'))^{-1}J(S - I)^{-1}. \tag{12.20}$$

We begin by remarking that the matrix $M_S = \frac{1}{2}J(S + I)(S - I)^{-1}$ is symmetric; this immediately follows from the conditions $S^T JS = J$. Notice that for every M with $\det(M - \frac{1}{2}J) \neq 0$ the equation

$$M = \tfrac{1}{2}J(S + I)(S - I)^{-1}$$

can be solved in S, yielding

$$S = (M - \tfrac{1}{2}J)^{-1}(M + \tfrac{1}{2}J);$$

the relation $S \in \mathrm{Sp}(n)$ is then equivalent to M being real and symmetric.

12.2.2 A Factorization Result

We have seen that every $\widehat{S} \in \mathrm{Mp}(n)$ can be written (non-uniquely) as the product $\widehat{S}_{W,m}\widehat{S}_{W',m'}$ of two metaplectic operators. We will need the following refinement of this property:

Lemma 8 *The generating quadratic forms W and W' in the factorizations $\widehat{S} = \widehat{S}_{W,m}\widehat{S}_{W',m'}$, $S = S_W S_{W'}$, can be chosen in such a way that* $\det[(S_W - I)(S_{W'} - I)] \neq 0$, *i.e.* $S_W, S_{W'} \in \mathrm{Sp}_0(n)$.

Proof We first remark that

$$\det(S_W - I) = \det(-B)\det\left[B^{-1}A + DB^{-1} - B^{-1} - (B^T)^{-1}\right] \tag{12.21}$$

(see Lemma 4 in de Gosson [5]). The next step consists in remarking that in view of the factorization (12.16) we have

$$\widehat{S}_{W,m}\widehat{S}_{W',m'} = \widehat{V}_{-P}\widehat{M}_{B^{-1},m}\widehat{J}\widehat{V}_{-(P'+Q)}\widehat{M}_{(B')^{-1},m'}\widehat{J}\widehat{V}_{-Q'} \tag{12.22}$$

with $P = DB^{-1}$, $P' = D'(B')^{-1}$, $Q = B^{-1}A$, $Q' = (B')^{-1}A'$. We next remark that the right hand-side of (12.22) remains unchanged if we replace P' with $P' + \lambda I$

and Q with $Q' - \lambda I$. Choosing λ so that

$$\det\left[P + Q - \lambda I + B^{-1} - (B^T)^{-1}\right] \neq 0$$
$$\det\left[P' + Q' + \lambda I - (B')^{-1} - (B'^T)^{-1}\right] \neq 0$$

we have $\det(S_W - I) \neq 0$ and $\det(S_{W'} - I) \neq 0$ in view of the equality (12.21). The lemma follows replacing, if necessary, $W(x, x')$ with $W(x, x') - \frac{1}{2}\lambda x^2$ and $W'(x, x')$ with $W'(x, x') + \frac{1}{2}\lambda x'^2$. ■

We remark that the symmetric matrix

$$W'' = B^{-1}A + DB^{-1} - B^{-1} - (B^T)^{-1} \tag{12.23}$$

is the Hessian matrix of the generating quadratic form W.

12.2.3 Metaplectic Operators as Weyl Operators

It turns out that metaplectic operators with projection $S \in \mathrm{Sp}_0(n)$ can be very simply represented by the Heisenberg operators:

Proposition 9 *Let $\widehat{S} \in \mathrm{Mp}(n)$ have projection $S \in \mathrm{Sp}_0(n)$. Then*

$$\widehat{S} = i^{\nu(\widehat{S})}\sqrt{|\det(S - I)|}\int \widehat{T}(Sz)\widehat{T}(-z)d^{2n}z \tag{12.24}$$

where the integer $\nu(\widehat{S})$ corresponds to the choice (modulo 2) of an argument of $\det(S - I)$*:*

$$\nu(\widehat{S})\pi \equiv n\pi + \arg\det(S - I) \mod 2\pi. \tag{12.25}$$

When $S = S_W$ is a free symplectic matrix we have

$$\nu(\widehat{S}_{W,m}) \equiv m - \mathrm{Inert}\, W'' \mod 2 \tag{12.26}$$

where Inert W'' *is the number of negative eigenvalues of the Hessian matrix* (12.23).

Proof See de Gosson [4, 6–8, 10]. ■

An alternative characterization is given by:

Proposition 10 *Let $S \in \mathrm{Sp}_0(n)$. The operator*

$$\widehat{R}_\nu(S) = \left(\frac{1}{2\pi\hbar}\right) i^\nu\sqrt{|\det(S - I)}\int \widehat{T}(Sz)\widehat{T}(-z)d^{2n}z \tag{12.27}$$

can be written in the two following ways:

$$\widehat{R}_\nu(S) = \left(\frac{1}{2\pi\hbar}\right)^n \frac{i^\nu}{\sqrt{|\det(S-I)|}} \int e^{\frac{i}{2\hbar}M_S z^2}\widehat{T}(z)d^{2n}z \tag{12.28}$$

and

$$\widehat{R}_\nu(S) = \left(\frac{1}{2\pi\hbar}\right)^n \frac{i^\nu}{\sqrt{|\det(S-I)}} \int e^{-\frac{i}{2\hbar}\sigma(Sz,z)}\widehat{T}((S-I)z)d^{2n}z. \tag{12.29}$$

Proof We have

$$\tfrac{1}{2}J(S+I)(S-I)^{-1} = \tfrac{1}{2}J + J(S-I)^{-1}$$

hence, in view of the antisymmetry of J,

$$M_S z^2 = \langle J(S-I)^{-1}z, z\rangle = \sigma((S-I)^{-1}z, z).$$

Performing the change of variables $z \longmapsto (S-I)^{-1}z$ we can rewrite the integral in the right-hand side of (12.28) as

$$\begin{aligned}\int e^{\frac{i}{2}M_S z^2}\widehat{T}(z)d^{2n}z &= \int e^{\frac{i}{2}\sigma(z,(S-I)z)}\widehat{T}((S-I)z)d^{2n}z \\ &= \int e^{-\frac{i}{2}\sigma(Sz,z)}\widehat{T}((S-I)z)d^{2n}z\end{aligned}$$

hence (12.27). Taking into account the relation (6.15) we have

$$\widehat{T}((S-I)z) = e^{-\frac{i}{2}\sigma(Sz,z)}\widehat{T}(Sz)\widehat{T}(-z)$$

and formula (12.29) follows. ■

12.3 The τ-Metaplectic Group $\mathrm{Mp}_\tau(n)$

Metaplectic operators are Weyl operators, and hence be represented as Shubin operators.

12.3.1 The Operators R_τ

We will need the following well-known generalization of the Fresnel formula (see e.g. Folland [2], Appendix A): let X be a real invertible matrix of dimension m; then:

$$\int e^{-2\pi iuv}e^{2\pi iXv^2}dv = |\det X|^{-1/2}e^{\frac{i\pi}{4}\operatorname{sign}X}e^{-i\pi X^{-1}u^2} \tag{12.30}$$

where sign X is the difference between the number of >0 and <0 eigenvalues of X.

Using this formula we set out to study the operators

$$R_\tau(S) = \sqrt{|\det(S-I)|}\int \widehat{T}_\tau(Sz)\widehat{T}_\tau(-z)d^{2n}z \tag{12.31}$$

defined for $S \in \mathrm{Sp}_0(n)$. They are continuous mappings $\mathcal{S}(\mathbb{R}^n) \longrightarrow \mathcal{S}(\mathbb{R}^n)$.

Proposition 11 *Let S, $S' \in \mathrm{Sp}_0(n)$ and assume that $SS' \in \mathrm{Sp}_0(n)$ as well. (i) We have*

$$R_\tau(SS') = e^{i\frac{\pi}{4}\operatorname{sign}M(SS')}R_\tau(S)R_\tau(S'). \tag{12.32}$$

(ii) The inverse of $R_\tau(S)^{-1}$ is given by the formula

$$R_\tau(S)^{-1} = R_\tau(S^{-1}) = R_{1-\tau}(S)^\dagger. \tag{12.33}$$

Proof (i) (Cf. the proof of Proposition 4.2 in de Gosson [8]). For brevity we write $M = M(S)$, $M' = M(S')$. In view of the composition formula for Weyl operators the twisted symbol c_σ of $R_\tau(S)R_\tau(S')$ is

$$c_\sigma(z) = K\int e^{\frac{i}{\hbar}[\sigma(z,z')+\phi(z,z')]}d^{2n}z'$$

where the constant K and the phase ϕ are given by

$$K = |\det(S-I)(S'-I)|^{-1/2}$$
$$\phi(z,z') = Mz^2 - 2Mz\cdot z' + (M+M')z'^2.$$

A straightforward calculation shows that

$$\sigma(z,z') - 2Mz\cdot z' = -2J(S-I)^{-1}z\cdot z'$$

hence

$$\sigma(z,z') + \phi(z,z') = -2J(S-I)^{-1}z\cdot z' + Mz^2 + (M+M')z'^2.$$

It follows that

$$c_\sigma(z) = Ke^{\frac{i}{\hbar}Mz^2}\int e^{-\frac{i}{\hbar}J(S-I)^{-1}z\cdot z'}e^{\frac{i}{\hbar}(M+M')z'^2}d^{2n}z'. \tag{12.34}$$

Applying the Fresnel formula (12.30) with $X = M + M'$ to (12.35) and replacing K with its value we get

$$c_\sigma(z) = |\det[(M + M')(S - I)(S' - I)]|^{-1/2} e^{\frac{i\pi}{4}\operatorname{sign}(M+M')} e^{\frac{i}{\hbar}\Theta(z)} \tag{12.35}$$

where the phase $\Theta(z)$ is given by

$$\begin{aligned}\Theta(z) &= \left[M + (S^T - I)^{-1} J (M + M')^{-1} J (S - I)^{-1}\right] z^2 \\ &= M(SS')z^2\end{aligned}$$

(the second equality in view of formula (12.20)). Noting that by definition of the symplectic Cayley transform we have

$$M + M' = J(I + (S - I)^{-1} + (S' - I)^{-1})$$

it follows, using again property (12.20), that

$$\begin{aligned}\det[(M + M')(S - I)(S' - I)] &= \det[(S - I)(M + M')(S' - I)] \\ &= \det(SS' - I)\end{aligned}$$

which concludes the proof of the product formula (12.32). (ii) Let us first show that $R_\tau(S^{-1}) = R_\tau(S)^{-1}$. Let c be the symbol of the operator $C = R_\tau(S)R_\tau(S^{-1})$; we claim that $c_\sigma(z) = \delta(z)$, hence $C = I$. Noting that $\det(S^{-1} - I) = \det(S - I) \neq 0$, formula (12.34) above shows that we have

$$c_\sigma(z) = Le^{\frac{i}{\hbar}Mz^2} \int e^{-\frac{i}{\hbar}J(S-I)^{-1}z\cdot z'} e^{\frac{i}{\hbar}(M+M(S^{-1}))z'^2} d^{2n}z'$$

where $L = |\det(S - I)|^{-1}$. Since $M(S^{-1}) = -M$ we have, setting $z'' = (S^T - I)^{-1}Jz'$,

$$\begin{aligned}c_\sigma(z) &= \frac{e^{\frac{i}{\hbar}Mz^2}}{|\det(S - I)|} \int e^{-\frac{i}{\hbar}J(S-I)^{-1}z\cdot z'} d^{2n}z' \\ &= e^{\frac{i}{\hbar}Mz^2} \int e^{\frac{i}{\hbar}zz''} d^{2n}z''\end{aligned}$$

hence $c_\sigma(z) = \delta(z)$ by the Fourier inversion formula, which proves that $C = I$. Let us finally show that $R_\tau(S^{-1}) = R_{1-\tau}(S)^\dagger$. We have

$$\begin{aligned}R_\tau(S^{-1}) &= \frac{1}{\sqrt{|\det(S^{-1} - I)|}} \int e^{\frac{i}{\hbar}M(S^{-1})z^2} \widehat{T}_\tau(z) d^{2n}z \\ &= \frac{1}{\sqrt{|\det(S - I)|}} \int e^{-\frac{i}{\hbar}M(S)z^2} \widehat{T}_\tau(z) d^{2n}z\end{aligned}$$

hence, using formula (9.33) for the adjoint of a τ-operator,

$$R_\tau(S^{-1})^\dagger = \frac{1}{\sqrt{|\det(S-I)|}} \int e^{\frac{i}{\hbar}M(S)z^2} \widehat{T}_{1-\tau}(z) d^{2n}z = R_{1-\tau}(S)$$

which is the same thing as $R_\tau(S^{-1}) = R_{1-\tau}(S)^\dagger$. ■

We make here an important remark: while the standard metaplectic operators $\widehat{S} = R_\tau(S) \in \mathrm{Mp}(n)$ are unitary, formula (12.33) shows that the operators $R_\tau(S)$ are not, unless $\tau = \frac{1}{2}$.

12.3.2 Definition of $\mathrm{Mp}_\tau(n)$

We now consider the group $\mathrm{Mp}_\tau(n)$ is generated by the τ-metaplectic operators $R_\tau^{\nu(S_W)}(S_W)$.

Definition 12 Let τ be an arbitrary real number. $\mathrm{Mp}_\tau(n)$ is the group of operators $\mathcal{S}(\mathbb{R}^n) \longrightarrow \mathcal{S}(\mathbb{R}^n)$ generated by the operators $R_\tau^{\nu(S_W)}(S_W)$.

For an arbitrary $S \in \mathrm{Sp}(n)$ the operator $R_\tau^\nu(S) \in \mathrm{Mp}_\tau(n)$ is thus the product of a finite number of $R_\tau^{\nu(S_W)}(S_W)$ with $S_W \in \mathrm{Sp}_0(n)$. We have moreover the following equivalent of the fact that each $\widehat{S} \in \mathrm{Mp}(n)$ can be written as the product of two metaplectic operators of the type $\widehat{S}_{W,m}$:

Proposition 13 *Let $S \in \mathrm{Sp}_0(n)$ and choose $S_W, S_{W'} \in \mathrm{Sp}_0(n)$ such that $S = S_W S_{W'}$. We have*

$$R_\tau^{\nu(S_W)}(S_W) R_\tau^{\nu(S_{W'})}(S_{W'}) = R_\tau^{\nu(S)}(S_W S_{W'}) \tag{12.36}$$

where the index $\nu(S)$ is given by the formula

$$\nu(S) = \nu(S_W) + \nu(S_{W'}) + \tfrac{1}{2}\operatorname{sign} M(SS'). \tag{12.37}$$

Proof It follows from formula (12.32) in Proposition 11 that we have

$$R_\tau^{\nu(S_W)}(S_W) R_\tau^{\nu(S_{W'})}(S_{W'}) = e^{i\phi} R_\tau(S) R_\tau(S')$$

$$\phi = \frac{\pi}{2}(\nu(S_W) + \nu(S_{W'}) + \tfrac{1}{2}\operatorname{sign} M(S_W S_{W'}))$$

hence the assertion. ■

References

1. V.C. Buslaev, Quantization and the W.K.B. method. Trudy Mat. Inst. Steklov. **110**, 5–28 (1978). (in Russian)
2. G.B. Folland, *Harmonic Analysis in Phase Space, Annals of Mathematics Studies* (Princeton University Press, Princeton, 1989)
3. M. de Gosson, Maslov indices on the metaplectic group Mp(n). Ann. Inst. Fourier **40**(3), 537–555 (1990)
4. M. de Gosson, *Maslov Classes, Metaplectic Representation and Lagrangian Quantization*, Mathematical Research 95 (Wiley VCH, Berlin, 1997)
5. M. de Gosson, The Weyl representation of metaplectic operators. Lett. Math. Phys. **72**, 129–142 (2005)
6. M. de Gosson, *Symplectic Geometry and Quantum Mechanics*. Operator Theory: Advances and Applications. Advances in Partial Differential Equations, vol. 166 (Birkhäuser, Basel, 2006)
7. M. de Gosson, S. de Gosson. An extension of the Conley–Zehnder index, a product formula and an application to the Weyl representation of metaplectic operators. J. Math. Phys. **47**(12), 123506, 15 (2006)
8. M. de Gosson, Metaplectic representation, Conley-Zehnder index, and Weyl calculus on phase space. Rev. Math. Phys. **19**(8), 1149–1188 (2007)
9. M. de Gosson, On the usefulness of an index due to Leray for studying the intersections of Lagrangian and symplectic paths. J. Math. Pures Appl. **91**, 598–613 (2009)
10. M. de Gosson, *Symplectic Methods in Harmonic Analysis and in Mathematical Physics* (Birkhäuser, Basel, 2011)
11. J. Leray, *Lagrangian Analysis and Quantum Mechanics*. A mathematical structure related to asymptotic expansions and the maslov index (the MIT Press, Cambridge, Mass., 1981); translated from Analyse Lagrangienne RCP 25, (Strasbourg Collège de France, 1976–1977)
12. V.P. Maslov, V.C. Bouslaev, V.I. Arnol'd, *Théorie des perturbations et méthodes asymptotiques* (Dunod, 1972)
13. H. Reiter, Metaplectic groups and Segal algebras. Lect. Notes Math. **1382** (1989)
14. E. Segal, Foundations of the theory of dynamical systems of infinitely many degrees of freedom. Mat. Fys. Medd. Danske Vid. Selsk **31**(12), 1–39 (1959)
15. D. Shale, Linear symmetries of free boson fields. Trans. Am. Math. Soc. **103**, 149–167 (1962)
16. L. van Hove, Sur certaines représentations unitaires d'un groupe fini de transformations. Mém. Acad. Roy. Belg. Classe des Sci. **26**(6), (1952)
17. N. Wallach, *Lie Groups: History, Frontiers and Applications*, vol. 5, Symplectic Geometry and Fourier Analysis (Math. Sci. Press, Brookline, 1977)
18. A. Weil, Sur certains groupes d'opérateurs unitaires. Acta Math. **111**, 143–211 (1964)

Chapter 13
Symplectic Covariance Properties

Given an operator $\widehat{A} = \text{Op}(a)$ a natural question that arises is what happens to that operator when one makes a change of variables in the symbol a. More precisely, can we relate in an easy way the operators $\widehat{A}$ and $\widehat{A}_\Phi = \text{Op}(a \circ \Phi)$ when Φ is a diffeomorphism of $\mathbb{R}^{2n}$? It turns out that this problem has no simple answer when one considers arbitrary operators and arbitrary diffeomorphisms, even in the linear case. However, if we restrict ourselves to Weyl pseudo-differential operators and linear symplectic transformations, then we have a simple "symplectic covariance property". We begin by reviewing this "easy" case, and thereafter study what remains of it when one replaces Weyl operators by Shubin and Born–Jordan operators.

13.1 Symplectic Covariance of Weyl Operators

A striking feature of Weyl calculus is that it is the *only* pseudo-differential operator calculus for which the property of symplectic covariance holds; that is, for every $\widehat{S} \in \text{Mp}(n)$ we have

$$\text{Op}_\text{W}(a \circ S) = \widehat{S}^{-1}\text{Op}_\text{W}(a)\widehat{S}$$

where $S = \pi(\widehat{S})$ is the projection of $\widehat{S}$ on the symplectic group $\text{Sp}(n)$.

13.1.1 *The Heisenberg and Grossmann–Royer Operators Revisited*

The key to all symplectic covariance results in Weyl pseudo-differential calculus is the following simple result about Heisenberg and Grossmann–Royer operators. We

M.A. de Gosson, *Born–Jordan Quantization*, Fundamental Theories of Physics 182, DOI 10.1007/978-3-319-27902-2_13

are going to give a rather short dynamical proof here; for a purely calculatory proof using the apparatus of generating functions see de Gosson [5], Sect. 8.1.3.

Proposition 1 *Let* $\widehat{S} \in \mathrm{Mp}(n)$ *and* $S = \pi(\widehat{S})$. *We have*

$$\widehat{S}\widehat{T}(z_0)\widehat{S}^{-1} = \widehat{T}(Sz_0) \tag{13.1}$$

$$\widehat{S}\widehat{T}_{GR}(z_0)\widehat{S}^{-1} = \widehat{T}_{GR}(Sz_0) \tag{13.2}$$

for every $z_0 \in \mathbb{R}^{2n}$.

Proof Recall from Chap. 6, Sect. 6.2.3, that the Heisenberg operator is the time-one solution of the Schrödinger equation

$$i\hbar\partial_t\psi = \widehat{H}_0\psi\,,\ \psi(\cdot,0) = \psi_0 \tag{13.3}$$

where $\widehat{H}_0$ is the displacement Hamiltonian $H_0(z) = \sigma(z, z_0)$ whose flow consists of the phase space translations $f_t : z \longmapsto z + tz_0$. Consider the unitary operator $U_t^S = \widehat{S}\widehat{T}(tz_0)\widehat{S}^{-1}$; the function $\psi = U_t^S\psi_0$ is the solution of the Schrödinger equation

$$i\hbar\frac{\partial\psi}{\partial t} = (\widehat{S}\widehat{H}_{z_0}\widehat{S}^{-1})\psi$$

with $\psi(\cdot, 0) = \psi_0$. Now, $\widehat{H}_0 = e^{-\frac{i}{\hbar}\sigma(\widehat{z},z_0)t}$ hence

$$\widehat{S}\widehat{H}_{z_0}\widehat{S}^{-1} = e^{-\frac{i}{\hbar}\widehat{S}\sigma(\widehat{z},z_0)\widehat{S}^{-1}t}$$

A direct calculation shows that

$$\widehat{S}\sigma(\widehat{z}, z_0)\widehat{S}^{-1} = \sigma(\widehat{z}, Sz_0)$$

hence we also have $\psi(x,t) = \widehat{T}(tSz_0)$. The equality (13.1) follows, taking $t = 1$. Formula (13.2) follows from (13.1) using the relation (6.12). In fact, denoting by $R^\vee$ the reflection operator $x \longmapsto -x$,

$$\begin{aligned}\widehat{S}\widehat{T}_{\mathrm{GR}}(z_0)\widehat{S}^{-1} &= \widehat{S}\widehat{T}(z_0)\widehat{S}^{-1}R^\vee\widehat{S}\widehat{T}(z_0)^{-1}\widehat{S}^{-1}\\ &= \widehat{T}(Sz_0)R^\vee\widehat{T}(Sz_0)^{-1}\\ &= \widehat{T}_{\mathrm{GR}}(Sz_0)\end{aligned}$$

(we have used here the obvious identity $\widehat{S}^{-1}R^\vee\widehat{S} = R^\vee$; see ([5], Chapter 8, Sect. 8.3.3). ∎

13.1.2 Weyl Operators

The following very important result easily follows from Proposition 1:

Proposition 2 *Let $S \in \mathrm{Sp}(n)$ and $\widehat{S} \in \mathrm{Mp}(n)$ such that $S = \pi(\widehat{S})$. For every $a \in \mathcal{S}'(\mathbb{R}^{2n})$ we have*

$$\mathrm{Op_W}(a \circ S) = \widehat{S}^{-1}\mathrm{Op_W}(a)\widehat{S}. \tag{13.4}$$

Proof We first observe that if $a \in \mathcal{S}'(\mathbb{R}^{2n})$ then $a \circ S \in \mathcal{S}'(\mathbb{R}^{2n})$; this is easily checked using the invariance of the Schwartz space $\mathcal{S}(\mathbb{R}^{2n})$ under linear changes of coordinates. Let us set $\widehat{B} = \mathrm{Op}(a \circ S)$. In view of formula (6.40) we have

$$\widehat{B}\psi = \int a_\sigma(Sz)\widehat{T}(z)\psi d^{2n}z$$

(the integral being understood as usual in the distributional sense). Performing the change of variables $Sz \longmapsto z$ and taking into account the fact that $\det S = 1$, we have

$$\widehat{B}\psi = \int a_\sigma(z)\widehat{T}(S^{-1}z)\psi d^{2n}z.$$

By formula (13.1) we have $\widehat{S}^{-1}\widehat{T}(z)\widehat{S} = \widehat{T}(S^{-1}z)$ and hence

$$\begin{aligned}\widehat{B}\psi &= \int a_\sigma(z)\widehat{S}^{-1}\widehat{T}(z)\widehat{S}\psi d^{2n}z \\ &= \widehat{S}^{-1}\left(\int a_\sigma(z)\widehat{T}(z)d^{2n}z\right)\widehat{S}\psi\end{aligned}$$

which is precisely (13.4). ∎

The following very important result shows that one cannot expect other quantizations to enjoy the full symplectic covariance property (13.4):

Proposition 3 *Let $a \longmapsto \mathrm{Op}(a)$ be a continuous linear mapping from $\mathcal{S}'(\mathbb{R}^{2n})$ to the space $\mathcal{L}(\mathcal{S}(\mathbb{R}^n), \mathcal{S}'(\mathbb{R}^n))$. Assume that:*

(i) if a only depends on $x \in \mathbb{R}^n$ and $a \in L^\infty(\mathbb{R}^n)$, then $\mathrm{Op}(a)$ is multiplication by $a(x)$;
(ii) if $S \in \mathrm{Sp}(n)$ then $\mathrm{Op}(a \circ S) = \widehat{S}\mathrm{Op}(a)\widehat{S}^{-1}$.

Then $a \longmapsto \mathrm{Op}(a)$ is the Weyl correspondence: $\mathrm{Op}(a) = \mathrm{Op_W}(a)$.

Proof The result was already stated in Stein [9], and rigorously proven in Wong [10], Chap. 30. ∎

The property of symplectic covariance thus really singles out the Weyl correspondence among all possible pseudo-differential calculi. We will see below that, however, some partial symplectic covariance survives for Shubin and Born–Jordan operators.

A straightforward consequence of Proposition 1 are the following formulas satisfied by the cross-Wigner and ambiguity functions:

Corollary 4 *Let* $\psi, \phi \in \mathcal{S}(\mathbb{R}^n)$ *and* $\widehat{S} \in \mathrm{Mp}(n)$*; We have*

$$\mathrm{Wig}(\widehat{S}\psi, \widehat{S}\phi)(z) = \mathrm{Wig}(\psi, \phi)(S^{-1}z) \tag{13.5}$$

$$\mathrm{Amb}(\widehat{S}\psi, \widehat{S}\phi)(z) = \mathrm{Amb}(\psi, \phi)(S^{-1}z). \tag{13.6}$$

where $S = \pi(\widehat{S})$. *and hence in particular*

$$\mathrm{Wig}(\widehat{S}\psi)(z) = \mathrm{Wig}\psi(S^{-1}z). \tag{13.7}$$

Proof In view of formula (7.19) in Proposition 8 we have

$$\langle\langle a, \mathrm{Wig}(\widehat{S}\psi, \widehat{S}\phi)\rangle\rangle = \langle \widehat{A}\widehat{S}\psi, (\widehat{S}\phi)^*\rangle$$

since $\widehat{S}$ is unitary we thus have

$$\int \mathrm{Wig}(\widehat{S}\psi, \widehat{S}\phi)(z)a(z)d^{2n}z = \langle \widehat{S}^{-1}\widehat{A}\widehat{S}\psi, \phi^*\rangle.$$

In view of (13.4) we have

$$\begin{aligned}\langle \widehat{S}^{-1}\widehat{A}\widehat{S}\psi, \phi^*\rangle &= \int \mathrm{Wig}(\psi, \phi)(z)(a \circ S)(z)d^{2n}z \\ &= \int \mathrm{Wig}(\psi, \phi)(S^{-1}z)a(z)d^{2n}z\end{aligned}$$

which establishes the equality (13.5) since ψ and ϕ are arbitrary. The formula (13.6) follows, since $\mathrm{Amb}(\psi, \phi)$ is the symplectic Fourier transform of $\mathrm{Wig}(\psi, \phi)$:

$$\begin{aligned}\mathrm{Amb}(\widehat{S}\psi, \widehat{S}\phi)(z) &= F_\sigma(\mathrm{Wig}(\widehat{S}\psi, \widehat{S}\phi))(z) \\ &= F_\sigma(\mathrm{Wig}(\psi, \phi) \circ S^{-1})(z) \\ &= F_\sigma(\mathrm{Wig}(\psi, \phi))(S^{-1}z) \\ &= \mathrm{Amb}(\psi, \phi)(S^{-1}z);\end{aligned}$$

to go from the second to the third line we used the relation $F_\sigma(a \circ S^{-1}) = F_\sigma a \circ S^{-1}$ which follows from the definition (6.21) of F_σ. ■

Note that both (13.7) and (13.6) can also be directly proven using the covariance formulas (13.1) and (13.2).

13.1.3 Affine Covariance Properties

Let us denote by $T(z_0)$ the phase space translation operator $z \longmapsto z+z_0$. It induces a natural action of functions by the formula $T(z_0)a(z) = a(z-z_0)$; this action can be extended into an action on distributions in the obvious way.

Proposition 5 *Let $\widehat{A} = \mathrm{Op_W}(a)$ with $a \in \mathcal{S}'(\mathbb{R}^{2n})$. Let $T(z_0)a(z) = a(z-z_0)$. We have*

$$\mathrm{Op_W}(T(z_0)a) = \widehat{T}(z_0)\mathrm{Op_W}(a)\widehat{T}(z_0)^{-1}. \tag{13.8}$$

Proof We begin by noting that

$$F_\sigma(T(z_0)a)(z) = e^{-\frac{i}{\hbar}\sigma(z,z_0)}a_\sigma(z)$$

as immediately follows from the definition

$$a_\sigma(z) = \left(\tfrac{1}{2\pi\hbar}\right)^n \int e^{-\frac{i}{\hbar}\sigma(z,z')}a(z')d^{2n}z'$$

of the symplectic Fourier transform and the bilinearity of the symplectic form σ. It follows

$$\mathrm{Op_W}(T(z_0)a) = \int a_\sigma(z)e^{-\frac{i}{\hbar}\sigma(z,z_0)}\widehat{T}(z)d^{2n}z.$$

In view of the commutation relation (6.14) in Chap. 6 we have

$$\widehat{T}(z)\widehat{T}(z_0) = e^{\frac{i}{\hbar}\sigma(z,z_0)}\widehat{T}(z_0)\widehat{T}(z)$$

and hence

$$e^{-\frac{i}{\hbar}\sigma(z,z_0)}\widehat{T}(z) = \widehat{T}(z_0)\widehat{T}(z)\widehat{T}(z_0)^{-1}$$

so that

$$\begin{aligned}\mathrm{Op_W}(T(z_0)a) &= \int a_\sigma(z)e^{-\frac{i}{\hbar}\sigma(z,z_0)}\widehat{T}(z_0)\widehat{T}(z)\widehat{T}(z_0)^{-1}d^{2n}z \\ &= \widehat{T}(z_0)\mathrm{Op_W}(T(z_0)a)\widehat{T}(z_0)^{-1}\end{aligned}$$

which is formula (13.8). ∎

Formula (13.8) can also be proven using the relation

$$\langle \widehat{A}\psi, \phi^*\rangle = \langle\langle a, W(\phi,\psi)\rangle\rangle$$

together with the covariance formula (13.5) for the cross-Wigner transform.

Formulas (13.4) and (13.8) are often summarized by saying that Weyl calculus is covariant with respect to the inhomogeneous metaplectic group IMp(n) (IMp(n) is the group of unitary operators generated by the metaplectic operators $\widehat{S} \in \mathrm{Mp}(n)$ and the Heisenberg operators $\widehat{T}(z_0)$, $z_0 \in \mathbb{R}^{2n}$; see [1, 2]).

13.2 The Case of τ-Operators

We are following rather closely our exposition in de Gosson [6].

13.2.1 A Covariance Result

Recall from the previous chapter that $\mathrm{Mp}_\tau(n)$ is the group generated by the operators

$$R_\tau(S) = \sqrt{|\det(S-I)|}\int \widehat{T}_\tau(Sz)\widehat{T}_\tau(-z)d^{2n}z \tag{13.9}$$

where $S \in \mathrm{Sp}_0(n)$ (the set of symplectic transformations with no eigenvalue equal to one), and the τ-Heisenberg operator

$$\widehat{T}_\tau(z_0) = e^{\frac{i}{2\hbar}(2\tau-1)p_0x_0}T(z_0). \tag{13.10}$$

The operators $R_\tau(S)$ are continuous linear mappings $\mathcal{S}(\mathbb{R}^n) \longrightarrow \mathcal{S}(\mathbb{R}^n)$, which are non-unitary for $\tau \neq \frac{1}{2}$.

The following result generalizes formula (13.1) to these operators:

Proposition 6 *Let $S \in \mathrm{Sp}_0(n)$.*

(i) We have the intertwining relation

$$R_\tau(S)\widehat{T}_\tau(z)R_\tau(S)^{-1} = \widehat{T}_\tau(Sz). \tag{13.11}$$

(ii) For every $R_\tau^\nu(S) \in \mathrm{Mp}_\tau(n)$ we have, for $a \in \mathcal{S}'(\mathbb{R}^{2n})$,

$$R_\tau(S)\mathrm{Op}_\tau(a)R_\tau(S)^{-1} = \mathrm{Op}_\tau(a \circ S^{-1}). \tag{13.12}$$

Proof (i) It is equivalent to show that the operators

$$\Gamma_\tau(S) = \int \widehat{T}_\tau(Sz)\widehat{T}_\tau(-z)d^{2n}z$$

satisfy the relation

$$\Gamma_\tau(S)\widehat{T}_\tau(z) = \widehat{T}_\tau(Sz)\Gamma_\tau(S). \tag{13.13}$$

Let $\psi \in \mathcal{S}(\mathbb{R}^n)$; in view of formula (9.13) we have

$$\Gamma_\tau(S)\psi = \int e^{\frac{i}{\hbar}\sigma(Sz,z)}\widehat{T}_\tau((S-I)z)\psi d^{2n}z;$$

since $S - I$ is a linear automorphism, $\widehat{T}_\tau((S-I)z) : \mathcal{S}(\mathbb{R}^n) \longrightarrow \mathcal{S}(\mathbb{R}^n)$ hence $\Gamma_\tau(S)\psi \in \mathcal{S}(\mathbb{R}^n)$. The continuity of $\Gamma_\tau(S)$ is straightforward to verify. Set

$$F(z,z_0) = \widehat{T}_\tau(Sz)\widehat{T}_\tau(-z)\widehat{T}_\tau(z_0)$$
$$G(z,z_0) = \widehat{T}_\tau(Sz_0)\widehat{T}_\tau(Sz)\widehat{T}_\tau(-z).$$

By repeated use of formula (9.13) we obtain

$$F(z,z_0) = e^{-\frac{i}{\hbar}\sigma(Sz-z_0,z-z_0)}\widehat{T}_\tau((S-I)z+z_0)$$
$$G(z,z_0) = e^{-\frac{i}{\hbar}\sigma((S-I)z_0+Sz_0,z)}\widehat{T}_\tau((S-I)z+Sz_0)$$

hence $G(z-z_0,z_0) = F(z,z_0)$. It follows that

$$\int F(z,z_0)d^{2n}z = \int G(z,z_0)d^{2n}z$$

hence the equality (13.13). (ii) That the operators $R_\tau(S)$ satisfy the intertwining relation (13.12) follows using the definition of $\mathrm{Op}_\tau(a)$. ■

13.2.2 Application to the τ-Wigner Function

As we have seen above the cross-Wigner function $W(\psi,\phi)$ enjoys the symplectic covariance property

$$W(\widehat{S}\psi,\widehat{S}\phi)(z) = W(\psi,\phi)(S^{-1}z). \tag{13.14}$$

In the τ-dependent case this result must be modified as follows:

Proposition 7 *Let $S \in \mathrm{Sp}_0(n)$ and $\psi,\phi \in \mathcal{S}(\mathbb{R}^n)$. We have*

$$W_\tau(R_\tau(S)\psi, R_{1-\tau}(S)\phi)(z) = W_\tau(\psi,\phi)(S^{-1}z). \tag{13.15}$$

Proof Let $A_\tau = \mathrm{Op}_\tau(a)$. Recall that

$$(\mathrm{Op}_\tau(a)\psi|\phi) = \langle\langle a|W_\tau(\psi,\phi)\rangle\rangle$$

(formula (9.34)). In view of the second equality (12.33) we have, using

$$
\begin{aligned}
(R_\tau(S)\mathrm{Op}_\tau(a)\psi|R_{1-\tau}(S)\phi) &= (R_{1-\tau}(S)^\dagger R_\tau(S)\mathrm{Op}_\tau(a)\psi|\phi) \\
&= (\mathrm{Op}_\tau(a)\psi|\phi) \\
&= \langle\langle a|W_\tau(\psi,\phi)\rangle\rangle.
\end{aligned}
$$

On the other hand, using the intertwining property (13.12), we have

$$
\begin{aligned}
(R_\tau(S)\mathrm{Op}_\tau(a)\psi|R_{1-\tau}(S)\phi) &= (\mathrm{Op}_\tau(a\circ S)R_\tau(S)\psi|R_{1-\tau}(S)\phi) \\
&= \langle a\circ S, W_\tau(R_\tau(S)\psi, R_{1-\tau}(S)\phi)\rangle \\
&= \langle a, W_\tau(R_\tau(S)\psi, R_{1-\tau}(S)\phi)\circ S^{-1}\rangle\rangle
\end{aligned}
$$

(the last identity using the change of variables $z \longmapsto S^{-1}z$ and the fact that $\det S = 1$). Formula (13.15) follows. ■

Notice that we recover the usual symplectic covariance formula (13.14) for the cross-Wigner transform when $\tau = \frac{1}{2}$.

It follows that our operators $R_\tau(S)$ coincide, up to an unimodular factor with metaplectic operators when $\tau = \frac{1}{2}$:

Corollary 8 *For $S \in \mathrm{Sp}_0(n)$ the operators $R(S) = R_{1/2}(S)$ are, up to a unimodular factor $i^{\nu(S)}$ elements of the metaplectic group* $\mathrm{Mp}(n)$.

Proof In [3, 4] we have shown that

$$
R^\nu(S) = \frac{i^{\nu(S)}}{\sqrt{|\det(S-I)|}} \int e^{\frac{i}{\hbar}M(S)z^2}\widehat{T}(z)d^{2n}z \tag{13.16}
$$

where $\nu(S)$ is the Conley–Zehnder index which we describe below. The result follows since $\widehat{T}_{1/2}(z) = \widehat{T}(z)$. ■

The Conley–Zehnder index $i_{\mathrm{CZ}}(\widetilde{S})$ is, loosely speaking, a Maslov-type intersection index measuring the number of intersections of a symplectic path $\widetilde{S} : t \longmapsto S_t$, $0 \leq t \leq 1$ joining the identity to $S \in \mathrm{Sp}_0(n)$ with the variety $\{S : \det(S-I) = 0\}$ which plays the role of a "caustic". More precisely, $i_{\mathrm{CZ}}(\widetilde{S})$ associates to the path $t \longmapsto S_t$ an integer which only depends on the homotopy class (with fixed endpoints) of that path, and such that

$$
\operatorname{sign}\det(S-I) = (-1)^{n-i_{\mathrm{CZ}}(\widetilde{S})}. \tag{13.17}
$$

Equivalently,

$$
\frac{1}{\pi}\arg\det(S-I) \equiv n - i_{\mathrm{CZ}}(\widetilde{S}) \bmod 2. \tag{13.18}
$$

In [7] we have proven that when S is a free symplectic matrix S_W then

$$i_{\mathrm{CZ}}(\widetilde{S_W}) \equiv m + n - \operatorname{Inert} W'' \bmod 2. \tag{13.19}$$

For a detailed discussion of that index (including an extension to the case det $(S - I) = 0$ and its expression in terms of the Leray–Maslov index [8] see de Gosson [2, 4].

In analogy with the definition of Mp(n) in terms of the basic metaplectic operators $\widehat{S}_{W,m}$ we define a group Mp$_\tau(n)$ of (non-unitary) operators as follows. Let $S_W \in \mathrm{Sp}_0(n)$ be a free symplectic matrix; we know from Lemma 8 in Chap. 12 that every $S \in \mathrm{Sp}(n)$ is the product of two such matrices. Let $\widehat{S}_{W,m} \in \mathrm{Mp}(n)$ be anyone of the two metaplectic operators covering S_W, and define

$$R_\tau^{\nu(S_W)}(S_W) = i^{\nu(S_W)} R_\tau(S_W) \tag{13.20}$$

$$\nu(S_W) \equiv n\pi + \arg\det(S - I) \bmod 2\pi. \tag{13.21}$$

The operators $R_\tau^{\nu(S_W)}(S_W)$ are invertible, and

$$R_\tau^{\nu(S_W)}(S_W)^{-1} = R_\tau^{-\nu(S_W)}(S_W^{-1}) = R_\tau^{-\nu(S_W)}(S_{W^*}) \tag{13.22}$$

(the first equality in view of the first formula (12.33) in Proposition 11, and the second since the free symplectic matrix S_W^{-1} is generated by $W^*(x, x') = -W(x', x)$ (formula (12.33) in Proposition 7).

13.3 The Case of Born–Jordan Operators

The intertwining properties for τ operators do not carry over to the Born–Jordan case; it is meaningless to expect a relation like $R_{\mathrm{BJ}}(S)\widehat{T}_{\mathrm{BJ}}(z) = \widehat{T}_{\mathrm{BJ}}(Sz)R_{\mathrm{BJ}}(S)$ which would lead to a full symplectic covariance property. However, as we have proven in [6] Born–Jordan operators enjoy a symplectic covariance property for operators belonging to a subgroup of the standard metaplectic group Mp(n).

13.3.1 The Born–Jordan Metaplectic Group

Recall from the previous chapter that the metaplectic group Mp(n) is generated by the modified Fourier transform $\widehat{J} = i^{-n/2}F$, the multiplication operators

$$\widehat{V}_{-P}\psi = e^{iPx^2/2\hbar}\psi \ (P = P^T)$$

and the unitary scaling operators

$$\widehat{M}_{L,m}\psi(x) = i^m \sqrt{|\det L|}\psi(Lx)$$

($\det L \neq 0$, $m\pi = \arg\det L$). The projections of these operators on $\mathrm{Sp}(n)$ are, respectively, the symplectic matrices J and

$$V_{-P} = \begin{pmatrix} I & 0 \\ P & I \end{pmatrix}, \; M_L = \begin{pmatrix} L^{-1} & 0 \\ 0 & L^2 \end{pmatrix}.$$

Proposition 9 *Let $\widehat{A}_{\mathrm{BJ}} = \mathrm{Op}_{\mathrm{BJ}}(a)$ with $a \in \mathcal{S}'(\mathbb{R}^{2n})$. We have*

$$\widehat{S}\mathrm{Op}_{\mathrm{BJ}}(a)\widehat{S}^{-1} = \mathrm{Op}_{\mathrm{BJ}}(a \circ S^{-1}) \tag{13.23}$$

for every $\widehat{S} \in \mathrm{Mp}(n)$ which is a product of a (finite number) of operators $\widehat{J}$ and $\widehat{M}_{L,m}$.

Proof It suffices to prove formula (13.23) for $\widehat{S} = \widehat{J}$ and $\widehat{S} = \widehat{M}_{L,m}$. Let first $\widehat{S}$ be an arbitrary element of $\mathrm{Mp}(n)$; we have

$$\begin{aligned} \widehat{S}\mathrm{Op}_{\mathrm{BJ}}(a) &= \left(\tfrac{1}{2\pi\hbar}\right)^n \int a_\sigma(z)\Theta(z)\widehat{S}\widehat{T}(z)d^{2n}z \\ &= \left[\left(\tfrac{1}{2\pi\hbar}\right)^n \int a_\sigma(z)\Theta(z)\widehat{T}(Sz)d^{2n}z\right]\widehat{S} \end{aligned}$$

where the second equality follows from the usual symplectic covariance property $\widehat{S}\widehat{T}(z) = \widehat{T}(Sz)\widehat{S}$ of the Heisenberg operators. Making the change of variables $z' = Sz$ in the integral we get, since $\det S = 1$,

$$\int a_\sigma(z)\Theta(z)\widehat{T}(Sz)d^{2n}z = \int a_\sigma(S^{-1}z)\Theta(S^{-1}z)\widehat{T}(z)d^{2n}z.$$

Now, by definition of the symplectic Fourier transform we have

$$\begin{aligned} a_\sigma(S^{-1}z) &= \left(\tfrac{1}{2\pi\hbar}\right)^n \int e^{-\frac{i}{\hbar}\sigma(S^{-1}z,z')}a(z')d^{2n}z' \\ &= (a \circ S^{-1})_\sigma(z). \end{aligned}$$

Choosing $\widehat{S} = \widehat{M}_{L,m}$ we have

$$\Theta(M_L^{-1}z) = \frac{\sin(2\pi Lp \cdot (L^T)^{-1}x)}{2\pi Lp \cdot (L^T)^{-1}x} = \Theta(z);$$

similarly $\Theta(J^{-1}z) = \Theta(z)$, hence in both cases

$$\begin{aligned}\widehat{S}\mathrm{Op}_{\mathrm{BJ}}(a) &= \left(\int (a \circ S^{-1})_\sigma \Theta(z)\widehat{T}(z)d^{2n}z\right)\widehat{S} \\ &= \mathrm{Op}_{\mathrm{BJ}}(a \circ S^{-1})\widehat{S}\end{aligned}$$

whence formula (13.23). ■

The proof above shows that the essential step consists in noting that $\Theta(S^{-1}z) = \Theta(z)$ when $S = J$ or $S = M_L$. It is clear that this property fails if one takes any operator $S = V_P$ with $P \neq 0$, so we cannot expect to have full symplectic covariance for Born–Jordan operators since the symplectic group $\mathrm{Sp}(n)$ is generated by the set of all matrices J, M_L and V_P. Full symplectic covariance is anyway excluded since this property is characteristic of Weyl calculus (Proposition 3), and can thus not be satisfied by any other pseudo-differential calculus.

This suggests the following definition:

Definition 10 The Born–Jordan metaplectic group is the subgroup $\mathrm{Mp}_{\mathrm{BJ}}(n)$ of $\mathrm{Mp}(n)$ generated by the Fourier transform $\widehat{J}$ and the unitary rescaling operators $\widehat{M}_{L,m}$.

The Born–Jordan metaplectic group is the maximal group of symmetries for Born–Jordan operators. The following easy result identifies $\mathrm{Mp}_{\mathrm{BJ}}(n)$:

Proposition 11 *The group* $\mathrm{Mp}_{\mathrm{BJ}}(n)$ *consists of all unitary operators of the type*

$$\widehat{S} = \widehat{M}_{L,m} \text{ or } \widehat{S} = \widehat{J}\widehat{M}_{L,m}$$

and $\pi(\mathrm{Mp}_{\mathrm{BJ}}(n))$ *is the subgroup of* $\mathrm{Sp}(n)$ *consisting of the symplectic matrices of the type*

$$M_L = \begin{pmatrix} L^{-1} & 0 \\ 0 & L^T \end{pmatrix} \text{ or } JM_L = \begin{pmatrix} 0 & L^T \\ -L^{-1} & 0 \end{pmatrix}.$$

Proof Every $\widehat{S} \in \mathrm{Mp}_{\mathrm{BJ}}(n)$ is a product of operators $\widehat{M}_{L,m}$ and $\widehat{J}$ hence $\pi(\mathrm{Mp}_{\mathrm{BJ}}(n))$ consists of all products of matrices M_L and JM_L. One easily verifies that every element of $\pi(\mathrm{Mp}_{\mathrm{BJ}}(n))$ can only be of the type M_L or JM_L. The result follows. ■

Notice that $\mathrm{Mp}_{\mathrm{BJ}}(n)$ contains the group of orthogonal transformations $O(n, \mathbb{R})$ identified with a subgroup of $O(2n, \mathbb{R})$ using the embedding

$$R \longrightarrow M_R = \begin{pmatrix} R & 0 \\ 0 & R \end{pmatrix}, R \in O(n, \mathbb{R}).$$

13.3.2 The Born–Jordan Wigner Transform Revisited

Recall that the Born–Jordan Wigner transform of $(\psi, \phi) \in \mathcal{S}(\mathbb{R}^n)$ is defined by

$$W_{\mathrm{BJ}}(\psi, \phi)(z) = \int_0^1 W_\tau(\psi, \phi)(z)dt. \tag{13.24}$$

In Proposition 5 in Chap. 10 we showed that $\widehat{A}_{\mathrm{BJ}} = \mathrm{Op}_{\mathrm{BJ}}(a)$ is related to $W_{\mathrm{BJ}}(\psi, \phi)$ by the formula

$$(\widehat{A}_{\mathrm{BJ}}\psi|\phi) = \langle\langle a, W_{\mathrm{BJ}}(\psi, \phi)\rangle\rangle. \tag{13.25}$$

This enables us to prove the following partial symplectic covariance result for the Born–Jordan Wigner transform:

Proposition 12 *Let $(\psi, \phi) \in \mathcal{S}(\mathbb{R}^n)$. For every $\widehat{S} \in \mathrm{Mp}_{\mathrm{BJ}}(n)$ with projection $S = \pi(\widehat{S})$ we have*

$$W_{\mathrm{BJ}}(\widehat{S}\psi, \widehat{S}\phi)(z) = W_{\mathrm{BJ}}(\psi, \phi)(S^{-1}z). \tag{13.26}$$

Proof We have

$$\langle\langle a, W_{\mathrm{BJ}}(\widehat{S}\psi, \widehat{S}\phi)\rangle\rangle = (\widehat{A}_{\mathrm{BJ}}(\widehat{S}\psi)|\widehat{S}\phi) = (\widehat{S}^{-1}\widehat{A}_{\mathrm{BJ}}(\widehat{S}\psi)|\phi).$$

Since $\widehat{S}^{-1}\widehat{A}_{\mathrm{BJ}}(\widehat{S}\psi) = \mathrm{Op}_{\mathrm{BJ}}(a \circ S)$ in view of the covariance formula (13.23) we thus have

$$\begin{aligned}\langle\langle a, W_{\mathrm{BJ}}(\widehat{S}\psi, \widehat{S}\phi)\rangle\rangle &= (\mathrm{Op}_{\mathrm{BJ}}(a \circ S)|\phi) \\ &= \langle\langle a \circ S, W_{\mathrm{BJ}}(\widehat{S}\psi, \widehat{S}\phi)\rangle\rangle \\ &= \langle\langle a, W_{\mathrm{BJ}}(\widehat{S}\psi, \widehat{S}\phi) \circ S^{-1}\rangle\rangle\end{aligned}$$

hence formula (13.26) since ψ and ϕ are arbitrary. ■

References

1. G.B. Folland, *Harmonic Analysis in Phase Space* (Princeton University Press, Princeton, 1989). Annals of Mathematics Studies
2. M. de Gosson, *Symplectic Geometry and Quantum Mechanics* (Birkhäuser, Basel, 2006)
3. M. de Gosson, Metaplectic representation, Conley–Zehnder index, and Weyl calculus on phase space. Rev. Math. Phys. **19**(8), 1149–1188 (2007)
4. M. de Gosson, On the usefulness of an index due to Leray forstudying the intersections of Lagrangian and symplectic paths. J. Math. Pures Appl. **91**, 598–613 (2009)
5. M. de Gosson, *Symplectic Methods in Harmonic Analysis* (Applications to Mathematical Physics, Birkhäuser, 2011)

6. M. de Gosson, Symplectic covariance properties for Shubin and Born–Jordan pseudo-differential operators. Trans. Am. Math. Soc. **365**, (2013)
7. K. Gröchenig, *Foundations of Time-Frequency Analysis* (Birkhäuser, Boston, 2000)
8. J. Leray, *Lagrangian Analysis and Quantum Mechanics, Amathematical Structure Related to Asymptotic Expansions and the Maslov Index* (The MIT Press, Cambridge, 1981); translated from *AnalyseLagrangienne* RCP 25, Strasbourg Collège de France, 1976–1977
9. E.M. Stein, *Harmonic Analysis: Real Variable Methods, Orthogonality, and Oscillatory Integrals* (Princeton University Press, 1993)
10. M.W. Wong, *Weyl Transforms* (Springer, 1998)

Chapter 14
Symbol Classes and Function Spaces

In this chapter we initiate the study of continuity properties for Born–Jordan operators. We will discuss the global symbol classes introduced by Shubin; they are "global" in the sense that they satisfy growth estimates with an equal weighting on the position and momentum variables. The use of these classes will allows us to considerably simplify the study of regularity properties of Born–Jordan pseudo-differential operators by reducing this study to the case of Shubin operators. If the choice of good symbol classes is essential in any pseudo-differential calculus, so is the choice of good functional spaces between which these operators act. These spaces must reflect regularity and continuity properties of the operators. We will discuss some of them here (a special emphasis being made on Feichtinger's modulation spaces).

14.1 Shubin's Symbol Classes

It is convenient—and natural in our context—to introduce Shubin's symbol classes (see Shubin [8], particularly §23). Their specificity lies in the fact that the Shubin symbols satisfy global estimates in the phase space variable $z = (x, p)$ (also see Nicola and Rodino [12]). This is in strong contrast with the usual Kohn–Nirenberg pseudo-differential calculus used in the theory of partial differential equations [2] or time-frequency analysis [10].

In what follows we will use the weight function $\langle\cdot\rangle$ on $\mathbb{R}^m$ defined by

$$\langle u \rangle = \sqrt{1 + |u|^2},\ u \in \mathbb{R}^m.$$

M.A. de Gosson, *Born–Jordan Quantization*, Fundamental Theories of Physics 182, DOI 10.1007/978-3-319-27902-2_14

14.1.1 The Shubin Symbol Classes Γ_ρ^m and $\Pi_\rho^{m,m'}$

We begin by defining good symbol classes; the following definition is due to Shubin [13]:

Definition 1 Let $m \in \mathbb{R}$ and $0 < \rho \leq 1$. The symbol class $\Gamma_\rho^m(\mathbb{R}^{2n})$ consists of all complex functions $a \in C^\infty(\mathbb{R}^{2n})$ such that for every $\alpha \in \mathbb{N}^{2n}$ there exists a constant $C_\alpha \geq 0$ with

$$|\partial_z^\alpha a(z)| \leq C_\alpha \langle z\rangle^{m-\rho|\alpha|} \text{ for } z \in \mathbb{R}^{2n}. \tag{14.1}$$

We set

$$\Gamma_\rho^{-\infty}(\mathbb{R}^{2n}) = \bigcap\nolimits_{m\in\mathbb{R}} \Gamma_\rho^m(\mathbb{R}^{2n}) = \mathcal{S}(\mathbb{R}^{2n}). \tag{14.2}$$

Obviously $\Gamma_\rho^m(\mathbb{R}^{2n})$ and $\Gamma_\rho^{-\infty}(\mathbb{R}^{2n})$ are complex vector spaces for the usual operations of addition and multiplication by complex numbers. Moreover one easily checks that

$$a \in \Gamma_\rho^m(\mathbb{R}^{2n}) \text{ and } b \in \Gamma_\rho^{m'}(\mathbb{R}^{2n}) \Longrightarrow ab \in \Gamma_\rho^{m+m'}(\mathbb{R}^{2n}) \tag{14.3}$$

$$a \in \Gamma_\rho^m(\mathbb{R}^{2n}) \text{ and } \alpha \in \mathbb{N}^{2n} \Longrightarrow \partial_z^\alpha a \in \Gamma_\rho^{m-|\alpha|}(\mathbb{R}^{2n}). \tag{14.4}$$

The first implication is proved by using the generalized Leibniz rule for the derivatives of a product of functions; the second is obvious in view of the definition of $\Gamma_\rho^m(\mathbb{R}^{2n})$.

Example 2 The reduced harmonic oscillator Hamiltonian $H(z) = \frac{1}{2}(|x|^2 + |p|^2)$ obviously belongs to $\Gamma_1^2(\mathbb{R}^{2n})$, and so does

$$H(z) = \sum_{j=1}^n \frac{1}{2m_j}(p_j^2 + m_j^2\omega_j^2 x_j^2).$$

More generally any polynomial function in z of degree m is in $\Gamma_1^m(\mathbb{R}^{2n})$.

The following result shows that the class $\Gamma_\rho^m(\mathbb{R}^{2n})$ is preserved by linear changes of variables (this property does not hold for the usual Hörmander classes, or their variants).

Proposition 3 *Let $a \in \Gamma_\rho^m(\mathbb{R}^{2n})$ and M a linear automorphism of $\mathbb{R}^{2n}$. Then $a \circ M \in \Gamma_\rho^m(\mathbb{R}^{2n})$.*

Proof Let us first show that $|a(Mz)| \leq C_M \langle z\rangle^m$ for some constant $C_M \geq 0$. Diagonalizing the symmetric automorphism $M^T M$ using an orthogonal transformation we have

$$\lambda_{\min}|z|^2 \leq |Mz|^2 \leq \lambda_{\max}|z|^2$$

where $\lambda_{\min} > 0$ and $\lambda_{\max} > 0$ are the smallest and largest eigenvalues of $M^T M$. It follows that

$$\langle Mz\rangle^m \leq \max(1, \lambda_{\max}) \langle z\rangle^m$$

if $m \geq 0$, and

$$\langle Mz\rangle^m \leq \min(1, \lambda_{\min}) \langle z\rangle^m$$

if $m < 0$. We thus have $|a(Mz)| \leq C_M \langle z\rangle^m$ for some constant $C_M > 0$. A similar argument shows that for every multi-index $\alpha \in \mathbb{N}^{2n}$ we have an estimate of the type

$$|\partial_z^\alpha (a \circ M)(z)| \leq C_{\alpha,M} \langle z\rangle^{m-\rho|\alpha|}$$

where $C_{\alpha,M} > 0$. ■

In practice we often work with operators of the type

$$\widehat{A}\psi(x) = \left(\tfrac{1}{2\pi\hbar}\right)^n \int e^{\frac{i}{\hbar}p(x-y)} a(x, y, p)\psi(y) d^n y d^n p; \qquad (14.5)$$

their symbols are defined, not on $\mathbb{R}^{2n} \equiv \mathbb{R}^n_x \times \mathbb{R}^n_p$ but rather on $\mathbb{R}^{3n} \equiv \mathbb{R}^n_x \times \mathbb{R}^n_y \times \mathbb{R}^n_p$. Here are a few examples:

Example 4 Shubin's τ-operators

$$\widehat{A}\psi(x) = \left(\tfrac{1}{2\pi\hbar}\right)^n \int e^{\frac{i}{\hbar}p(x-y)} a((1-\tau)x + \tau y)\psi(y) d^n y d^n p$$

and generalized Feynman operators

$$\widehat{A}\psi(x) = \left(\tfrac{1}{2\pi\hbar}\right)^n \int e^{\frac{i}{\hbar}p(x-y)} \tfrac{1}{2}(a(x, p) + a(y, p))\psi(y) d^n y d^n p$$

are of the type above. Born–Jordan operators $\widehat{A}_{\mathrm{BJ}} = \mathrm{Op}_{\mathrm{BJ}}(a)$ are also of the type (14.5) with

$$a_{\mathrm{BJ}}(x, y, p) = \int_0^1 a((1-\tau)x + \tau y) d\tau.$$

It therefore makes sense to define a symbol class generalizing $\Gamma^m_\rho(\mathbb{R}^{2n})$ by allowing a dependence on the three sets of variables x, y, and p.

Definition 5 Let $(m, m') \in \mathbb{R}^2$. The symbol class $\Pi^{m,m'}_\rho(\mathbb{R}^{3n})$ consists of all functions $a \in C^\infty(\mathbb{R}^{3n})$ having the following property: for every $(\alpha, \beta, \gamma) \in \mathbb{N}^{3n}$ there exists a constant $C_{\alpha\beta\gamma} \geq 0$ such that:

$$|\partial_p^\alpha \partial_x^\beta \partial_y^\gamma a(x, y, p)| \leq C_{\alpha\beta\gamma} \langle u\rangle^{m-\rho|\alpha+\beta+\gamma|} \langle x - y\rangle^{m'+\rho|\alpha+\beta+\gamma|} \qquad (14.6)$$

where $u = (x, y, p)$. The class of all operators (14.5) with $a \in \Pi_\rho^{m,m'}(\mathbb{R}^{3n})$ is denoted by $G_\rho^m(\mathbb{R}^n)$. We set $\Pi_\rho^{m,0}(\mathbb{R}^{3n}) = \Pi_\rho^m(\mathbb{R}^{3n})$.

The consideration of the symbol class $\Pi_\rho^{m,m'}(\mathbb{R}^{3n})$ will allow us to prove some deep results about Born–Jordan operators.

We will see below that every $\widehat{A} \in G_\rho^m(\mathbb{R}^n)$ is a τ-pseudo-differential operator. We will use several times Peetre's inequality

$$(1 + |\xi - \eta|^2)^s \leq 2^{|s|}(1 + |\xi|^2)^s(1 + |\eta|^2)^{|s|} \tag{14.7}$$

valid for $\xi, \eta \in \mathbb{R}^m$ and $s \in \mathbb{R}$ (see for instance Chazarain and Piriou [2]).

14.1.2 Asymptotic Expansions of Symbols

Let us now briefly study the notion of asymptotic expansion of a symbol $a \in \Gamma_\rho^m(\mathbb{R}^{2n})$; we are following closely the presentation in de Gosson [8], §14.1.3; for details see Shubin's treatise [13].

Definition 6 Let $(a_j)_j$ be a sequence of symbols $a_j \in \Gamma_\rho^{m_j}(\mathbb{R}^{2n})$ such that $\lim_{j\to+\infty} m \to -\infty$. Let $a \in C^\infty(\mathbb{R}^{2n})$. If for every integer $r \geq 2$ we have

$$a - \sum_{j=1}^{r-1} a_j \in \Gamma_\rho^{\overline{m}_r}(\mathbb{R}^{2n}) \tag{14.8}$$

where $\overline{m}_r = \max_{j\geq r} m_j$ we will write $a \sim \sum_{j=1}^\infty a_j$ and call this relation an asymptotic expansion of the symbol a.

The interest of the asymptotic expansion comes from the fact that every sequence of symbols $(a_j)_j$ with $a_j \in \Gamma_\rho^{m_j}(\mathbb{R}^{2n})$, the degrees m_j being strictly decreasing and such that $m_j \to -\infty$ determines a symbol in some $\Gamma_\rho^m(\mathbb{R}^{2n})$, that symbol being unique up to an element of $\mathcal{S}(\mathbb{R}^{2n})$:

Proposition 7 *Let $(a_j)_j$ be a sequence of symbols $a_j \in \Gamma_\rho^{m_j}(\mathbb{R}^{2n})$ such that $m_j > m_{j+1}$ and $\lim_{j\to+\infty} m \to -\infty$. Then:*

(i) There exists a function a, such that $a \sim \sum_{j=1}^\infty a_j$.

(ii) If another function a' is such that $a' \sim \sum_{j=1}^\infty a_j$, then $a - a' \in \mathcal{S}(\mathbb{R}^{2n})$.

Note that property (ii) immediately follows from the fact that we have

$$\bigcap_{m\in\mathbb{R}} \Gamma_\rho^m(\mathbb{R}^{2n}) = \mathcal{S}(\mathbb{R}^{2n})$$

14.1.3 A Reduction Result

It turns out that an operator (14.5) with symbol $a \in \Pi_{\rho}^{m,m'}(\mathbb{R}^{3n})$ is a Shubin τ-pseudo-differential operator—and this for every value of the parameter τ! To prove this important result we will need the following simple technical result:

Lemma 8 *Let f be a linear map $\mathbb{R}^{2n} \to \mathbb{R}^n$, and assume that the linear map $\phi : \mathbb{R}^{2n} \to \mathbb{R}^{2n}$ defined by $\phi(x, y) = (f(x, y), x - y)$ is an isomorphism. Let $b \in \Gamma_{\rho}^{m}(\mathbb{R}^{2n})$. The function a defined by the formula*

$$a(x, y, p) = b(f(x, y), p) \tag{14.9}$$

is in the symbol class $\Pi_{\rho}^{m,|m|}(\mathbb{R}^{3n})$.

Proof It is clear that $a \in C^{\infty}(\mathbb{R}^{3n})$. The functions $|x| + |y|$ and $|f(x, y)| + |x - y|$ give equivalent norms on $\mathbb{R}^{2n}$. Using Peetre's inequality (14.7) one easily shows that that

$$\frac{(1 + |f(x, y)| + |p|)^s}{(1 + |f(x, y)| + |x - y| + |p|)^s} \leq C(1 + |x - y|)^{|s|} \tag{14.10}$$

for all $s \in \mathbb{R}$. The estimates (14.6) follow for $a(x, y, p)$ with $m' = |m|$. ∎

Proposition 9 *Let τ be an arbitrary real number. (i) Every pseudo-differential operator $\widehat{A}$ of the type (14.5) with symbol $a \in \Pi_{\rho}^{m,m'}(\mathbb{R}^{3n})$ can be uniquely written in the form $\widehat{A} = \operatorname{Op}_{\tau}(a_{\tau})$ for some symbol $a_{\tau} \in \Gamma_{\rho}^{m}(\mathbb{R}^{2n})$, that is*

$$\widehat{A}\psi(x) = \left(\tfrac{1}{2\pi\hbar}\right)^n \int e^{\frac{i}{\hbar}p(x-y)} a_{\tau}((1-\tau)x + \tau y)\psi(y)d^n y d^n p;$$

the symbol a_{τ} has the asymptotic expansion

$$a_{\tau}(x, p) \sim \sum_{\beta,\gamma} \frac{1}{\beta!\gamma!} \tau^{|\beta|}(1-\tau)^{|\gamma|}$$
$$\partial_p^{\beta+\gamma}(i\hbar\partial_x)^{\beta}(-i\hbar\partial_y)^{\gamma} a(x, y, p)|_{y=x}.$$

(ii) In particular, choosing $\tau = \frac{1}{2}$, there exists $a_{\mathrm{W}} \in \Gamma_{\rho}^{m}(\mathbb{R}^{2n})$ such that $\widehat{A} = \operatorname{Op}_{\mathrm{W}}(a_{\mathrm{W}})$.

Proof We are following with a few minor modifications Shubin's original proof ([13], Theorem 23.2; also see the presentation in de Gosson [8], §14.5). For notational simplicity we assume that $\hbar = 1$. Let us set $v = (1 - \tau)x + \tau y$ and $w = x - y$, that is, equivalently,

$$x = v + \tau w, \; y = v - (1 - \tau)\, w. \tag{14.11}$$

The symbol a can then be written

$$a(x, y, p) = a(v + \tau w, v - (1 - \tau)\, w, p). \tag{14.12}$$

Expanding the right-hand side of (14.12) in a Taylor series at $w = 0$, we get $a = a_N + r_N$ where

$$a_N(x, y, p) = \sum_{|\beta+\gamma| \le N-1} \frac{(-1)^{|\gamma|}}{\beta!\gamma!} \tau^{|\beta|} (1 - \tau)^{|\gamma|} (x - y)^{\beta+\gamma} (\partial_x^\beta \partial_y^\gamma a)(v, v, p) \tag{14.13}$$

and the remainder term r_N is given by the formula

$$r_N(x, y, p) = \sum_{|\beta+\gamma|=N} c_{\beta\gamma}(x - y)^{\beta+\gamma} I_{\beta\gamma}(x, y, p) \tag{14.14}$$

the $c_{\beta\gamma}$ being constants, and

$$I_{\beta\gamma}(x, y, p) = \int_0^1 (1 - t)^{N-1} (\partial_x^\beta \partial_y^\gamma a)(v + t\tau w, v - t(1 - \tau)w, p)dt. \tag{14.15}$$

In (14.13) the expression $(\partial_x^\beta \partial_y^\gamma a)(v, v, p)$ signifies that we have replaced x and y with $v = (1 - \tau)\, x + \tau y$ in the expression $\partial_x^\beta \partial_y^\gamma a(x, y, p)$. The expression

$$(\partial_x^\beta \partial_y^\gamma a)(v + t\tau w, v - t(1 - \tau)w, p)$$

in (14.15) should be understood in a similar way. We next note that the operator with symbol

$$a_{\beta\gamma}(x, y, p) = (x - y)^{\beta+\gamma} (\partial_x^\beta \partial_y^\gamma a)(v, v, p)$$

is the same as the one with symbol

$$b_{\beta\gamma}(x, y, p) = i^{|\beta|+|\gamma|} (\partial_p^{\beta+\gamma} \partial_x^\beta \partial_y^\gamma a)(v, v, p)$$

as is immediately seen using partial integrations and the relation

$$(x - y)^{\beta+\gamma} e^{ip(x-y)} = i^{-(|\beta|+|\gamma|)} \partial_p^{\beta+\gamma} e^{ip(x-y)}.$$

It follows from formula (14.13) that we have $\widehat{A} = \widehat{A}_N + \widehat{R}_N$ where $\widehat{A}_N$ is the operator with τ-symbol

$$a_N(x, p) = \sum_{|\beta+\gamma| \le N-1} \frac{i^{|\beta|+|\gamma|}}{\beta!\gamma!} (-\tau)^{|\beta|} (1 - \tau)^{|\gamma|} \partial_p^{\beta+\gamma} \partial_x^\beta \partial_y^\gamma a(x, y, p)|_{y=x}$$

and $\widehat{R}_N$ is the operator with symbol r_N given by (14.14). Note that the operator $\widehat{R}_N$ is a linear combination of a finite number of terms having symbols of the type

$$\int_0^1 (\partial_p^{\beta+\gamma}\partial_x^{\beta}\partial_y^{\gamma}a)(v+t\tau w, v-t(1-\tau)w, p)(1-t)^{N-1}dt \tag{14.16}$$

with $|\beta+\gamma| = N$. Let us now show that $r_N \in \Pi_\rho^{m-2N\rho}(\mathbb{R}^{3n})$. For this it suffices to show that this is true for the integrand in (14.16), with all estimates uniform in t (note that this is obvious for each fixed $t \neq 0$ and true for $t = 0$ by Lemma 8). Using the trivial relations

$$v = (1-\tau)(v+t\tau w) + \tau(v-t(1-\tau)w),$$
$$tw = (v+t\tau w) - (v-t(1-\tau)w)$$

it is easy to see that there exists a constant $C > 0$ independent of $t \in [0,1]$ such that

$$C^{-1} \leqq \frac{|v+t\tau w| + |v-t(1-\tau)w|}{|v|+|tw|} \leqq C$$

and we thus have the estimate

$$\begin{aligned}&|(\partial_p^{\beta+\gamma}\partial_x^{\beta}\partial_y^{\beta}a)(v+t\tau w, v-t(1-\tau)w, p)| \\ &\leq C(1+|v|+|tv|+|p|)^{m-2\rho N}(1+|tw|)^{m'+2\rho N}.\end{aligned}$$

Since for $m'+2\rho N \geq 0$ we have the inequality

$$\begin{aligned}&(1+|tw|)^{m'+2\rho N} \\ &\leq (1+|v|+|tv|+|p|)^{m'+2\rho N}(1+|v|+|p|)^{-(m'+2\rho N)},\end{aligned}$$

it is clear that if, in addition, $m+m' \geqq 0$ and $m-2\rho N \leqq 0$, then

$$\begin{aligned}&|(\partial_p^{\beta+\gamma}\partial_x^{\beta}\partial_y^{\beta}a)(v+t\tau w, v-t(1-\tau)w, p)| \\ &\leq C'(1+|v|+|p|)^{-m'-2\rho N}(1+|v|+|tw|+|p|)^{m'+m} \\ &\leq C'(1+|v|+|p|)^{m-2\rho N}(1+|w|)^{m'+m} \\ &\leq C'(1+|v|+|w|+|p|)^{m-2\rho N}(1+|w|)^{m'+2m+2\rho N}\end{aligned}$$

where C' is independent of t. One estimates the derivatives in a similar way. Now, let the symbol $b' \in \Gamma_\varrho^m(\mathbb{R}^{2n})$ be such that

$$b'(x,p) \sim \sum_{N=0}^{\infty}(b_N(x,p) - b_{N-1}(x,p)).$$

Then, if $\widehat{A}'$ has τ-symbol $b'(x,p)$ it is clear that the kernel of the operator $\widehat{A} - \widehat{A}'$ is in $\mathcal{S}(\mathbb{R}^n \times \mathbb{R}^n)$. ■

14.1.4 First Continuity Results

The following result shows that the symbol classes $\Pi_\rho^{m,m'}(\mathbb{R}^{3n})$ lead to a neat operator calculus; writing as above formally an operator $\widehat{A}$ as

$$\widehat{A}\psi(x) = \left(\tfrac{1}{2\pi\hbar}\right)^n \int e^{\frac{i}{\hbar}p(x-y)} a(x,y,p)\psi(y)d^n y d^n p$$

we have:

Proposition 10 *(i) Every operator $\widehat{A} \in G_\rho^m(\mathbb{R}^n)$ is a continuous operator $\mathcal{S}(\mathbb{R}^n) \longrightarrow \mathcal{S}(\mathbb{R}^n)$ and can hence be extended into a continuous operator $\mathcal{S}'(\mathbb{R}^n) \longrightarrow \mathcal{S}'(\mathbb{R}^n)$. (ii) If $\widehat{A} \in G_\rho^m(\mathbb{R}^n)$, then $\widehat{A}^\dagger \in G_\rho^m(\mathbb{R}^n)$. (iii) If $\widehat{A} \in G_\rho^m(\mathbb{R}^n)$ and $\widehat{B} \in G_\rho^{m'}(\mathbb{R}^n)$ then $\widehat{A}\widehat{B} \in G_\rho^{m+m'}(\mathbb{R}^n)$.*

Proof (i) That we have a continuous extension $\widehat{A} : \mathcal{S}'(\mathbb{R}^n) \longrightarrow \mathcal{S}'(\mathbb{R}^n)$ follows from the first statement by duality provided that the transpose $\widehat{A^T}$ is also in $G_\rho^m(\mathbb{R}^n)$. This immediately follows from formula (9.44) in Proposition 17, using the estimates (14.1) satisfied by the Shubin symbols. The proof of the continuity property $\widehat{A} : \mathcal{S}(\mathbb{R}^n) \longrightarrow \mathcal{S}(\mathbb{R}^n)$ is classical, and similar to the one one uses to prove the continuity of operators with symbols in the Hörmander classes; we omit it here. (ii) follows from formula (9.45) Proposition 17. The proof of (iii) is omitted; see Shubin [13], Theorem 23.6. ■

The following boundedness results is important:

Proposition 11 *(i) Let $a \in \Gamma_\rho^0(\mathbb{R}^{2n})$. For every $\tau \in \mathbb{R}$ the Shubin operator $\widehat{A}_\tau = \mathrm{Op}_\tau(a)$ can be extended into a bounded operator on $L^2(\mathbb{R}^n)$. (ii) Let $a \in \Gamma_\rho^m(\mathbb{R}^{2n})$, $m < 0$. Then $\widehat{A}_\tau = \mathrm{Op}_\tau(a)$ can be extended into a compact operator on $L^2(\mathbb{R}^n)$.*

Proof The proof of both properties requires the anti-Wick formalism, which we do not study here. See Shubin [13], Theorem 24.3 for property (i) and Shubin [13], Theorem 24.4 for (ii). ■

In ordinary pseudo-differential calculus one often expresses continuity results in terms of the usual Sobolev spaces $H^s(\mathbb{R}^n)$, defined by the condition

$$\int |\widehat{\psi}(p)|^2(1+|p|^2)^s d^n p < \infty.$$

Since the vocation of the operators studied in this chapter is to incorporate global behavior, it is appropriate to introduce the following variant of these spaces:

Definition 12 For $s \in \mathbb{R}$ the global Sobolev space $Q^s(\mathbb{R}^n)$ consists of all $\psi \in \mathcal{S}'(\mathbb{R}^n)$ such that

$$Q^s(\mathbb{R}^n) = L^2_s(\mathbb{R}^{2n}) \cap H^s(\mathbb{R}^{2n}).$$

The norm on $Q^s(\mathbb{R}^n)$ is defined by $||\psi||_{Q^s} = ||L_s\psi||$ where

$$L_s\psi(x) = \left(\tfrac{1}{2\pi\hbar}\right)^{n/2} \int e^{\frac{i}{\hbar}px} \langle z\rangle^{s/2}\widehat{\psi}(p)d^n p.$$

The space $Q^s(\mathbb{R}^n)$ can be equipped with an inner product making it into a Hilbert space, and we have the equalities

$$\bigcap_{s\in\mathbb{R}} Q^s(\mathbb{R}^n) = \mathcal{S}(\mathbb{R}^n), \bigcup_{s\in\mathbb{R}} Q^s(\mathbb{R}^n) = \mathcal{S}'(\mathbb{R}^n) \tag{14.17}$$

and that following regularity result holds:

Proposition 13 *Every operator $\widehat{A} \in G^m_\rho(\mathbb{R}^n)$ is a continuous operator $Q^s(\mathbb{R}^n) \longrightarrow Q^{s-m}(\mathbb{R}^n)$.*

In particular, using (14.17) we have $\widehat{A} : \mathcal{S}(\mathbb{R}^n) \longrightarrow \mathcal{S}(\mathbb{R}^n)$ and $A : \mathcal{S}'(\mathbb{R}^n) \longrightarrow \mathcal{S}'(\mathbb{R}^n)$ (cf. Proposition 10, (i)).

We mention that the study of $Q^s(\mathbb{R}^n)$ is best understood within the framework of Feichtinger's modulation spaces which will be introduced later in this chapter.

14.2 Modulation Spaces

Feichtinger's [4–7] modulation spaces form a category of functional spaces which plays a fundamental role in many theoretical and practical questions in analysis. They were originally designed to study phase-space concentration problems in time-frequency analysis, but their importance in quantum mechanics has been more recently realized. In this section we review a subcategory; for a rather exhaustive treatment see Gröchenig [10]; in de Gosson [8] (Chap. 17) we have given a review of modulation spaces from the point of view of the Wigner formalism; we are following this exposition here. We refer to Feichtinger's review paper [7] for a comprehensive description of recent research and advances in the topic.

14.2.1 The Modulation Spaces M^q_s

Recall that $\langle z\rangle^s = (1+|z|^2)^{s/2}$. We denote by $L^q_s(\mathbb{R}^{2n})$ the space of all $\psi \in \mathcal{S}'(\mathbb{R}^{2n})$ such that $\langle\cdot\rangle^s\,\psi \in L^q(\mathbb{R}^{2n})$; it is equipped with the norm

$$||\psi||_{L^q_s} = \left(\int |\langle z \rangle \psi(z)|^q d^{2n} z \right)^{1/q}.$$

Definition 14 The modulation space $M^q_s(\mathbb{R}^n)$ consists of all $\psi \in \mathcal{S}'(\mathbb{R}^n)$ such that $\text{Wig}(\psi, \phi) \in L^q_s(\mathbb{R}^{2n})$ for every "window" $\phi \in \mathcal{S}(\mathbb{R}^n)$. When $q = 1$ and $s = 0$ the space $M^1_0(\mathbb{R}^n) = S_0(\mathbb{R}^n)$ is called the Feichtinger algebra.

One proves that $S_0(\mathbb{R}^n)$ is an algebra both for pointwise product and for convolution, and that one has the inclusions

$$\mathcal{S}(\mathbb{R}^n) \subset S_0(\mathbb{R}^n) \subset C^0(\mathbb{R}^n) \cap L^1(\mathbb{R}^n) \cap L^2(\mathbb{R}^n). \tag{14.18}$$

The modulation space $M^q_s(\mathbb{R}^n)$ is equipped with the family of norms defined by

$$||\psi||_{\phi, M^q_s(\mathbb{R}^n)} = ||\text{Wig}(\psi, \phi)||_{L^q_s} = \int |\text{Wig}(\psi, \phi)(z)|^q \langle z \rangle^s d^{2n} z. \tag{14.19}$$

We have defined $M^q_s(\mathbb{R}^n)$ by requiring that $W(\psi, \phi) \in L^q_s(\mathbb{R}^{2n})$ for every window ϕ. In fact, it suffices to verify this condition for one window. The two following results summarize the main properties of $M^q_s(\mathbb{R}^n)$. The first is about the topological properties of that space:

Proposition 15 *(i) We have $\psi \in M^q_s(\mathbb{R}^n)$ if and only if we have* $\text{Wig}(\psi, \phi) \in L^q_s(\mathbb{R}^{2n})$ *for one (and hence every) window $\phi \in \mathcal{S}(\mathbb{R}^n)$. The topology of $M^q_s(\mathbb{R}^n)$ is defined by using a single norm $|| \cdot ||_{\phi, M^q_s}$; moreover all the norms obtained by letting ϕ vary are equivalent. (ii) The modulation space $M^q_s(\mathbb{R}^n)$ is a Banach space for the topology defined by the norm $|| \cdot ||_{\phi, M^q_s}$. (iii) The Schwartz space $\mathcal{S}(\mathbb{R}^n)$ is a dense subspace of each of the modulation spaces $M^q_s(\mathbb{R}^n)$.*

Proof See Gröchenig [10], Chapter 11, especially Proposition 11.3.4. ■

The second result show that $M^q_s(\mathbb{R}^n)$ is invariant under translations and metaplectic transformations:

Proposition 16 *(i) The space $M^q_s(\mathbb{R}^n)$ is invariant under the action of the Heisenberg operators $\widehat{T}(z)$; in fact there exists a constant $C > 0$ such that*

$$||\widehat{T}(z)\psi||_{\phi, M^q_s} \leq C \langle z \rangle^s ||\psi||_{\phi, M^q_s}. \tag{14.20}$$

(ii) Let $\widehat{S} \in \text{Mp}(n)$. We have $\widehat{S}\psi \in M^q_s(\mathbb{R}^n)$ if and only if $\psi \in M^q_s(\mathbb{R}^n)$. In particular $M^q_s(\mathbb{R}^n)$ is invariant under the Fourier transform.

Proof (i) We can choose $\widehat{T}(z_0)\phi$ as a window since $\widehat{T}(z_0)\phi \in \mathcal{S}(\mathbb{R}^n)$ if and only if $\phi \in \mathcal{S}(\mathbb{R}^n)$. We have

$$\text{Wig}(\widehat{T}(z_0)\psi, \widehat{T}(z_0)\phi)(z) = \text{Wig}(\psi, \phi)(z - z_0)$$

hence it suffices to show that $L^q_s(\mathbb{R}^{2n})$ is invariant under the phase space translations $T(z_0) : z \longmapsto z + z_0$. We have

$$\begin{aligned}||T(z_0)\psi||^q_{L^q_s} &= \int |\psi(z - z_0)|^q \langle z\rangle^{qs} d^{2n}z \\ &= \int |\psi(z)|^q \langle z + z_0\rangle^{qs} d^{2n}z \\ &\leq \langle z\rangle^s \int |\psi(z)|^q \langle z\rangle^{qs} d^{2n}z\end{aligned}$$

hence our claim; the estimate (14.20) follows. (ii) We have $\psi \in M^q_s(\mathbb{R}^n)$ if and only if $\mathrm{Wig}(\psi, \phi) \in L^q_s(\mathbb{R}^{2n})$ for one window $\phi \in \mathcal{S}(\mathbb{R}^n)$; if this property holds, then it holds for all windows. In view of the symplectic covariance of the Wigner transform we have

$$\begin{aligned}\mathrm{Wig}(\widehat{S}\psi, \phi) &= \mathrm{Wig}(\widehat{S}\psi, \widehat{S}(\widehat{S}^{-1}\phi))(z) \\ &= \mathrm{Wig}(\psi, (\widehat{S}^{-1}\phi))(S^{-1}z)\end{aligned}$$

hence $\mathrm{Wig}(\widehat{S}\psi, \phi) \in L^q_s(\mathbb{R}^{2n})$ if and only if the function $\mathrm{Wig}(\psi, (\widehat{S}^{-1}\phi)) \circ S^{-1}$ is in $L^q_s(\mathbb{R}^{2n})$. But this condition is equivalent to $\mathrm{Wig}(\psi, (\widehat{S}^{-1}\phi)) \in L^q_s(\mathbb{R}^{2n})$ since $\widehat{S}^{-1}\phi$ can be chosen as a window, hence $\widehat{S}\psi \in M^q_s(\mathbb{R}^n)$. ∎

This result in particular applies to the Feichtinger algebra $S_0(\mathbb{R}^n) = M^1_0(\mathbb{R}^n)$. In fact, one can show (Gröchenig [10]) that $S_0(\mathbb{R}^n)$ is the smallest Banach algebra containing the space $\mathcal{S}(\mathbb{R}^n)$ which is invariant under the action of the metaplectic group $\mathrm{Mp}(n)$ and the Heisenberg operators.

The Shubin spaces $Q^s(\mathbb{R}^n)$ are particular cases of modulation spaces:

Proposition 17 *We have $Q^s(\mathbb{R}^n) = M^2_s(\mathbb{R}^n)$ for every $s \in \mathbb{R}$.*

Proof See [1], Lemma 2.3. ∎

14.2.2 The Generalized Sjöstrand Classes

Let us set, for $s \geq 0$,

$$\langle\langle(z, \zeta)\rangle\rangle^s = (1 + |z|^2 + |\zeta|^2)^{s/2}. \tag{14.21}$$

Definition 18 The modulation space $M^{\infty,1}_s(\mathbb{R}^{2n})$ consists of all $a \in \mathcal{S}'(\mathbb{R}^{2n})$ such that there exists a function $\phi \in \mathcal{S}(\mathbb{R}^{2n})$ for which

$$\int \sup_{z\in\mathbb{R}^{2n}} |\widetilde{\mathrm{Wig}}(a, \phi)(z, \zeta)| \langle\langle(z, \zeta)\rangle\rangle^s d\zeta < \infty \tag{14.22}$$

where $\widetilde{\mathrm{Wig}}$ is the cross-Wigner transform on $\mathbb{R}^{2n}$. One calls $M^{\infty,1}(\mathbb{R}^{2n}) = M_0^{\infty,1}(\mathbb{R}^{2n})$ the Sjöstrand class.

The space $M^{\infty,1}(\mathbb{R}^{2n})$ was introduced by Sjöstrand [14, 15] using very different methods. It is easy to check that for every window $\phi \in \mathcal{S}(\mathbb{R}^{2n})$ the formula

$$||a||^{\phi}_{M_s^{\infty,1}} = \int \sup_z \left[|\widetilde{\mathrm{Wig}}(a,\phi)(z,\zeta)| \langle\langle (z,\zeta)\rangle\rangle^s \right] d\zeta < \infty \tag{14.23}$$

defines a norm on $M_s^{\infty,1}(\mathbb{R}^{2n})$. As is the case for the modulation spaces $M_s^q(\mathbb{R}^n)$ condition (14.23) is independent of the choice of window ϕ, and when ϕ runs through $\mathcal{S}(\mathbb{R}^{2n})$ the functions $||\cdot||^{\phi}_{M_s^{\infty,1}}$ form a family of equivalent norms on $M_s^{\infty,1}(\mathbb{R}^{2n})$. It turns out that $M_s^{\infty,1}(\mathbb{R}^{2n})$ is a Banach space for the topology defined by any of these norms; moreover the Schwartz space $\mathcal{S}(\mathbb{R}^{2n})$ is dense in $M_s^{\infty,1}(\mathbb{R}^{2n})$.

The Sjöstrand classes $M_s^{\infty,1}(\mathbb{R}^{2n})$ contain many of the usual pseudo-differential symbol classes and we have the important inclusion

$$C_b^{2n+1}(\mathbb{R}^{2n}) \subset M_0^{\infty,1}(\mathbb{R}^{2n}) \tag{14.24}$$

where $C_b^{2n+1}(\mathbb{R}^{2n})$ is the vector space of all functions which are differentiable up to order $2n+1$ with bounded derivatives. In fact, for every window ϕ there exists a constant $C_\phi > 0$ such that

$$||a||^{\phi}_{M_s^{\infty,1}} \le C_\phi ||a||_{C^{2n+1}} = C_\phi \sum_{|\alpha|\le 2n+1} ||\partial_z^\alpha A||_\infty. \tag{14.25}$$

The generalized Sjöstrand classes are invariant under linear changes of variables:

Proposition 19 *Let M be a real invertible $2n \times 2n$ matrix. If $a \in M_s^{\infty,1}(\mathbb{R}^{2n})$ then $a \circ M \in M_s^{\infty,1}(\mathbb{R}^{2n})$, and there exists a constant $C_M > 0$ such that for every window ϕ and every $a \in M_s^{\infty,1}(\mathbb{R}^{2n})$ we have*

$$||a \circ M||^{\phi}_{M_s^{\infty,1}} \le C_M ||a||^{\psi}_{M_s^{\infty,1}} \tag{14.26}$$

where $\psi = \phi \circ M^{-1}$.

For a proof, see Proposition 7 in de Gosson and Luef [9].

We are next going to show that $M_s^{\infty,1}(\mathbb{R}^{2n})$ is invariant under the action of the metaplectic group $\mathrm{Mp}(2n)$ corresponding to the symplectic group $\mathrm{Sp}(2n)$ of the symplectic space $(\mathbb{R}^{2n} \times \mathbb{R}^{2n}, \sigma \oplus \sigma)$.

Proposition 20 *Let $\widetilde{S} \in \mathrm{Mp}(2n)$ and $a \in \mathcal{S}'(\mathbb{R}^{2n})$. We have $a \in M_s^{\infty,1}(\mathbb{R}^{2n})$ if and only if $\widetilde{S}a \in M_s^{\infty,1}(\mathbb{R}^{2n})$ and we have*

$$||\widetilde{S}a||^{\widetilde{S}\phi}_{M_s^{\infty,1}} \le \lambda^s_{\max} ||Sa||^{\phi}_{M_s^{\infty,1}} \tag{14.27}$$

where $\lambda^s_{\max}$ *is the largest eigenvalue of* $S^T S \in \mathrm{Sp}(2n)$, $S = \Pi(\widetilde{S})$.

Proof Let $S \in \mathrm{Sp}(2n)$ be the projection of $\widetilde{S} \in \mathrm{Mp}(2n)$. We have, using the symplectic covariance of the Wigner transform,

$$\begin{aligned} \|\widetilde{S}a\|^{\widetilde{S}\phi}_{M_s^{\infty,1}} &= \int \sup_{z \in \mathbb{R}^{2n}} \left[|\widetilde{\mathrm{Wig}(\widetilde{S}a, \widetilde{S}\phi)}(z,\zeta)| \langle\langle (z,\zeta) \rangle\rangle^s \right] d\zeta \\ &= \int \sup_{z \in \mathbb{R}^{2n}} \left[|\widetilde{\mathrm{Wig}(a, \phi)}(S^{-1}(z,\zeta))| \langle\langle (z,\zeta) \rangle\rangle^s \right] d\zeta \\ &= \int \sup_{z \in \mathbb{R}^{2n}} \left[|\widetilde{\mathrm{Wig}(a, \phi)}(z,\zeta))| \langle\langle S(z,\zeta) \rangle\rangle^s \right] d\zeta. \end{aligned}$$

Now $\langle\langle S(z,\zeta) \rangle\rangle \leq \lambda_{\max} \langle\langle z,\zeta \rangle\rangle$ hence

$$\|\widetilde{S}a\|^{\widetilde{S}\phi}_{M_s^{\infty,1}} \leq \lambda^s_{\max} \int \sup_{z \in \mathbb{R}^{2n}} \left[|\widetilde{\mathrm{Wig}(a, \phi)}(z,\zeta))| \langle\langle z,\zeta \rangle\rangle^s \right] d\zeta$$

which is the inequality (14.27). ∎

The following result (Gröchenig [11]) shows that the Weyl correspondence $a \overset{\mathrm{Weyl}}{\longleftrightarrow} A$ is a continuous mapping $M_s^{\infty,1}(\mathbb{R}^{2n}) \longrightarrow M_s^q(\mathbb{R}^n)$:

Proposition 21 *Let* $a \in M_s^{\infty,1}(\mathbb{R}^{2n})$. *The Weyl operator* $\widehat{A}_{\mathrm{W}} = \mathrm{Op}_{\mathrm{W}}(a)$ *is bounded on* $M_s^q(\mathbb{R}^n)$ *for every* $q \in [1,\infty]$, *and there exists a constant* $C > 0$ *independent of* q *such that following uniform estimate holds*

$$\|\widehat{A}_{\mathrm{W}}\|^{\mathrm{Op}}_{M_s^q} \leq C\|a\|_{M_s^{\infty,1}}$$

for all $a \in M_s^{\infty,1}(\mathbb{R}^{2n})$ ($\|\cdot\|^{\mathrm{Op}}_{M_s^q}$ *is the operator norm on the Banach space* $M_s^q(\mathbb{R}^n)$).

The Sjöstrand class $M^{\infty,1}(\mathbb{R}^{2n})$ contains the Hörmander symbol class $S^0_{0,0}(\mathbb{R}^{2n})$. The result above implies as a particular case a Calderón–Vaillancourt type result: if $a \in S^0_{0,0}(\mathbb{R}^{2n})$ then $\widehat{A} = \mathrm{Op}_{\mathrm{W}}(a)$ is bounded on $L^2(\mathbb{R}^n)$.

For our purposes the following property is very important:

Proposition 22 *Let* $a, b \in M_s^{\infty,1}(\mathbb{R}^{2n})$. *Then* $c = a \star_\hbar b \in M_s^{\infty,1}(\mathbb{R}^{2n})$. *In particular, for every window of the type* $\phi = \mathrm{Wig}\varphi$ *where* $\varphi \in \mathcal{S}(\mathbb{R}^n)$, *there exists a constant* $C_\phi > 0$ *such that*

$$\|a \star_\hbar b\|^{\phi}_{M_s^{\infty,1}} \leq C_\phi \|a\|^{\phi}_{M_s^{\infty,1}} \|b\|^{\phi}_{M_s^{\infty,1}}.$$

Since obviously $a^* \in M_s^{\infty,1}(\mathbb{R}^{2n})$ if and only and $a \in M_s^{\infty,1}(\mathbb{R}^{2n})$ the property above can be restated by saying that $M_s^{\infty,1}(\mathbb{R}^{2n})$ is a Banach $*$-algebra with respect to the Moyal product $\star_\hbar$ and the involution $a \longmapsto a^*$.

14.3 Applications to Born–Jordan Operators

We are going to prove the central result of this chapter, Proposition 24. It says that if the symbol a belongs to the Shubin class $\Pi_\rho^m(\mathbb{R}^{2n})$ then the Born–Jordan operator $\widehat{A}_{\mathrm{BJ}} = \mathrm{Op}_{\mathrm{BJ}}(a)$ is a Weyl operator with symbol also in $\Pi_\rho^m(\mathbb{R}^{2n})$.

14.3.1 A Fundamental Result

We are now going to show that every Born–Jordan operator with symbol in one of the Shubin classes is a Weyl operator with symbol in the same class; for this we will need the following elementary Lemma (F. Nicola):

Lemma 23 *Let ξ and η be positive numbers and $m \in \mathbb{R}$. We have*

$$\min\{\xi^m, \eta^m\} \leq C(\xi + \eta)^m \tag{14.28}$$

where $C = 1$, when $m \leq 0$ and $C = 2^{-m}$ when $m \geq 0$.

Proof The case $m \geq 0$ is straightforward: we have

$$\min\{\xi^m, \eta^m\} \leq \xi^m + \eta^m \leq (\xi + \eta)^m.$$

Suppose $m < 0$; if $\xi \leq \eta$ we have

$$\min\{\xi^m, \eta^m\} = \eta^m \leq 2^{-m}(\xi + \eta)^m;$$

the case $\xi > \eta$ follows in the same way. ∎

Proposition 24 *Let $\widehat{A}_{\mathrm{BJ}} = \mathrm{Op}_{\mathrm{BJ}}(a)$ with $a \in \Pi_\rho^m(\mathbb{R}^{2n})$. (i) For every $\tau \in \mathbb{R}$ there exists $a_\tau \in \Gamma_\rho^m(\mathbb{R}^{2n})$ such that $\widehat{A}_{\mathrm{BJ}} = \mathrm{Op}_\tau(a_\tau)$. (ii) In particular, every Born–Jordan operator with symbol $a \in \Pi_\rho^m(\mathbb{R}^{2n})$ is a Weyl operator with symbol $a_{\mathrm{W}} \in \Pi_\rho^m(\mathbb{R}^{2n})$.*

Proof Property (ii) follows from (i) choosing $\tau = \frac{1}{2}$. (ii) Consider the τ-pseudo-differential operator $A_\tau = \mathrm{Op}_\tau(a)$:

$$A_\tau \psi(x) = \left(\tfrac{1}{2\pi\hbar}\right)^n \int e^{\frac{i}{\hbar}p(x-y)} a((1-\tau)x + \tau y, p)\psi(y) d^n y d^n p$$

and set

$$a_{\mathrm{BJ}}(x, y, p) = \int_0^1 a((1-\tau)x + \tau y, p) d\tau. \tag{14.29}$$

We thus have

$$A_{\mathrm{BJ}}\psi(x) = \left(\tfrac{1}{2\pi\hbar}\right)^n \int e^{\frac{i}{\hbar}p(x-y)} a_{\mathrm{BJ}}(x,y,p)\psi(y)d^n y d^n p$$

which is of the type (14.5). Let us show that $a_{\mathrm{BJ}} \in \Pi_\rho^{m,m'}(\mathbb{R}^{3n})$, i.e. that we have estimates of the type

$$|\partial_x^\alpha \partial_y^\beta \partial_p^\gamma a_{\mathrm{BJ}}(x,y,p)| \leq C_{\alpha,\beta\gamma}\langle (x,y,p)\rangle^{m-\rho|\alpha+\beta+\gamma|}\langle x-y\rangle^{m'+\rho|\alpha+\beta+\gamma|} \tag{14.30}$$

for some $m' \in \mathbb{R}$ independent of α, β, γ. The result will follow using Proposition 9. Let us set

$$b_\tau(x,y,p) = a((1-\tau)x + \tau y, p);$$

we have

$$\partial_x^\alpha \partial_y^\beta \partial_p^\gamma b_\tau(x,y,p) = (1-\tau)^{|\alpha|}\tau^{|\beta|}\partial_x^{\alpha+\beta}\partial_p^\gamma b_\tau(x,y,p)$$

hence, since $a \in \prod_\rho^{m,m'}(\mathbb{R}^{3n})$, we have the estimates

$$\begin{aligned} &|\partial_x^\alpha \partial_y^\beta \partial_p^\gamma b_\tau(x,y,p)| \\ &\leq C_{\alpha+\beta,\gamma}(1-\tau)^{|\alpha|}\tau^{|\beta|}\langle((1-\tau)x+\tau y,p)\rangle^{m-\rho|\alpha+\beta+\gamma|}. \end{aligned} \tag{14.31}$$

Now, by Peetre's inequality (14.7) we have

$$\begin{aligned} &\langle((1-\tau)x+\tau y,p)\rangle^{m-\rho|\alpha+\beta+\gamma|} \\ &\leq C\langle (x,p)\rangle^{m-\rho|\alpha+\beta+\gamma|}\langle \tau(x-y)\rangle^{|m|+\rho|\alpha+\beta+\gamma|} \end{aligned}$$

as well as

$$\begin{aligned} &\langle((1-\tau)x+\tau y,p)\rangle^{m-\rho|\alpha+\beta+\gamma|} \\ &\leq C\langle (y,p)\rangle^{m-\rho|\alpha+\beta+\gamma|}\langle (1-\tau)(x-y)\rangle^{|m|+\rho|\alpha+\beta+\gamma|} \end{aligned}$$

Combining these two inequalities we find

$$\begin{aligned} &\langle((1-\tau)x+\tau y,p)\rangle^{m-\rho|\alpha+\beta+\gamma|} \\ &\leq C\min\{\langle (x,p)\rangle^{m-\rho|\alpha+\beta+\gamma|}, \langle (y,p)\rangle^{m-\rho|\alpha+\beta+\gamma|}\}\langle x-y\rangle^{|m|+\rho|\alpha+\beta+\gamma|} \end{aligned}$$

This implies that

$$\begin{aligned} &\langle((1-\tau)x+\tau y,p)\rangle^{m-\rho|\alpha+\beta+\gamma|} \\ &\leq C^{\prime m-\rho|\alpha+\beta+\gamma|}\langle x-y\rangle^{|m|+\rho|\alpha+\beta+\gamma|} \end{aligned}$$

where we used the inequality (14.28). Together with (14.31) this inequality implies (14.30) with $m' = |m|$ after an integration on τ. ∎

Example 25 Let $n = 1$ and choose for $a(x, p)$ the monomial $x^2 p^2$. We have

$$\operatorname{Op}_{\mathrm{BJ}}(x^2 p^2) = \tfrac{1}{3}(\widehat{x}^2\widehat{p}^2 + \widehat{p}\widehat{x}\widehat{p} + \widehat{p}^2\widehat{x}^2)$$

and

$$\operatorname{Op}_\tau(x^2 p^2) = \tau^2\widehat{x}^2\widehat{p}^2 + 2\tau(1-\tau)\widehat{p}\widehat{x}\widehat{p} + (1-\tau)^2\widehat{p}^2\widehat{x}^2).$$

An immediate calculation shows that

$$\operatorname{Op}_{\mathrm{BJ}}(x^2 p^2) = \operatorname{Op}_{1/2}(\tfrac{4}{3}x^2 p^2).$$

14.3.2 Application: Some Boundedness Results

The following consequence of Proposition 24 is straightforward:

Proposition 26 *Suppose that* $a \in \Gamma_\rho^m(\mathbb{R}^n)$. *Then the Born–Jordan operator* $\widehat{A}_{\mathrm{BJ}} = \operatorname{Op}_{\mathrm{BJ}}(a)$ *is a continuous operator* $\widehat{A}_{\mathrm{BJ}} : Q^s(\mathbb{R}^n) \longrightarrow Q^{s-m}(\mathbb{R}^n)$.

Proof In view of Proposition 24 we can write $\widehat{A}_{\mathrm{BJ}} = \operatorname{Op}_\tau(a_\tau)$ with $a_\tau \in \Gamma_\rho^m(\mathbb{R}^n)$. It now suffices to apply Proposition 13. ∎

We also have the following L^2-boundedness result:

Proposition 27 *Let* $a \in \Gamma_\rho^0(\mathbb{R}^{2n})$. *Then* $\widehat{A}_{\mathrm{BJ}} = \operatorname{Op}_{\mathrm{BJ}}(a)$ *is a bounded operator on* $L^2(\mathbb{R}^n)$.

Proof It suffices to apply Proposition 24 to Proposition 11, (i). ∎

Here is a rather general result, which shows the flexibility of Proposition 24; recall that $C_b^{2n+1}(\mathbb{R}^{2n})$ is the algebra of all complex functions a on $\mathbb{R}^{2n}$ whose derivatives $\partial_z^\alpha a$ are bounded for $|\alpha| \leq 2n+1$.

Proposition 28 *Let* $a \in C_b^{2n+1}(\mathbb{R}^{2n})$. *The operator* $\widehat{A}_{\mathrm{BJ}} = \operatorname{Op}_{\mathrm{BJ}}(a)$ *is bounded on the modulation spaces* $M^q(\mathbb{R}^n) = M_0^q(\mathbb{R}^n)$ *for every* $q \geq 1$.

Proof Recall (formula (14.24)) that $C_b^{2n+1}(\mathbb{R}^{2n}) \subset M_0^{\infty,1}(\mathbb{R}^{2n})$. It follows from Proposition 24 that there exists a symbol $b \in M_0^{\infty,1}(\mathbb{R}^{2n})$ such that $\widehat{A}_{\mathrm{BJ}} = \operatorname{Op}_{\mathrm{W}}(b)$. It now suffices to apply Proposition 21. ∎

We finally mention the following continuity result, which is a particular case of a more general statement due to Elena Cordero (unpublished[1]):

---

[1]Private communication.

Proposition 29 *Let* $a \in M^q(\mathbb{R}^{2n}) = M_0^q(\mathbb{R}^{2n})$ *and* $\widehat{A}_{\mathrm{BJ}} = \mathrm{Op}_{\mathrm{BJ}}(a)$. *For* $r \geq 1$ *and* $q \geq 2$ *the operator* $\widehat{A}_{\mathrm{BJ}}$ *extends into a bounded operator on* $M^r(\mathbb{R}^n) = M_0^r(\mathbb{R}^n)$ *and there exists a constant* $C_{q,r} > 0$ *such that*

$$||\widehat{A}_{\mathrm{BJ}}\psi||_{M^r} \leq C_{q,r}||a||_{M^q}\,||\psi||_{M^r}$$

for all $\psi \in M^r(\mathbb{R}^{2n})$.

We omit the proof; it is a particular case of previous results obtained by Cordero and Nicola [3] in the Weyl case.

References

1. P. Boggiatto, E. Cordero, K. Gröchenig, Generalized anti-Wick operators with symbols in distributional Sobolev spaces. Integr. Eqn. Oper. Theory **48**(4), 427–442 (2004)
2. J. Chazarain, A. Piriou, *Introduction à la théorie des équations aux dérivées partielles linéaires* (Gauthier–Villars, Paris, 1981). English translation: *Introduction to the theory of linear partial differential equations. Studies in Mathematics and its Applications*, vol. 14 (North–Holland, 1982)
3. E. Cordero, F. Nicola, Pseudo differential operators on L^p. Wiener Amalgam Modulation Spaces Int. Math. Res. Not. **2010**(10), 1860–1893 (2010)
4. H.G. Feichtinger, Un espace de Banach de distributions tempérées sur les groupes localement compact abéliens, C. R. Acad. Sci. Paris, Série A–B **290**(17), A791–A794 (1980)
5. H.G. Feichtinger, On a new Segal algebra. Monatsh. Math. **92**(4), 269–289 (1981)
6. H.G. Feichtinger, Banach spaces of distributions of Wiener's type and interpolation, in *Functional Analysis and Approximation* (Oberwohlfach, 1980). Internat. Ser. Numer. Math. vol. 60 (Birkhäuser, Basel, 1981), pp. 153–165
7. H.G. Feichtinger, Modulation spaces: looking back and ahead. Sampling Theory Signal Image Process. **5**(2), 109–140 (2006)
8. M. de Gosson, *Symplectic Methods in Harmonic Analysis and in Mathematical Physics* (Birkhäuser, Basel, 2011)
9. M. de Gosson, F. Luef, On the usefulness of modulation spaces in deformation quantization. J. Phys A: Math. Theor. **42**(1), 315205 (2009)
10. K. Gröchenig, *Foundations of Time-Frequency Analysis* (Birkhäuser, Boston, 2000)
11. K. Gröchenig, Time-frequency analysis on Sjöstrand's class. Rev. Mat. Iberoamericana **22**(2), 703–724 (2006)
12. F. Nicola, L. Rodino, *Global Pseudo-differential Calculus on Euclidean Spaces* Pseudo-Differential Operators (Birkhäuser, 2010)
13. M.A. Shubin, *Pseudodifferential Operators and Spectral Theory* (Springer, 1987) (original Russian edition in Nauka, Moskva 1978)
14. J. Sjöstrand, An algebra of pseudo-differential operators. Math. Res. Lett. **1**(2), 185–192 (1994)
15. J. Sjöstrand, *Wiener type algebras of pseudo-differential operators*, Séminaire sur les Équations aux Dérivées Partielles École Polytech., Palaiseau, Exp. No. IV, 21, (1994–1995)

References

R. Abraham, J.E. Marsden, *Foundations of Mechanics*, 2nd edn. (The Benjamin/Cummings Publishing Company, 1978)

G.S. Agarwal, E. Wolf, Calculus for functions of noncommuting operators and general phase-space methods in quantum mechanics I. Mapping theorems and ordering of functions of noncommuting operators. Phys. Rev. D **2**(10), 2161–2164 (1970)

Y. Aharonov, P.G. Bergmann, J. Lebowitz, Time symmetry in the quantum process of measurement. Phys. Rev. B **134**, B1410–B1416 (1964)

Y. Aharonov, D.Z. Albert, L. Vaidman, How the result of a measurement of a component of the spin of a spin-$\frac{1}{2}$ particle can turn out to be 100. Phys. Rev. Lett. **60**(14), 1351–1354 (1988)

Y. Aharonov, A. Botero, Quantum averages of weak values. Phys. Rev. A **72**, 052111 (2005)

Y. Aharonov, S. Popescu, J. Tollaksen, A time-symmetric formulation of quantum mechanics. Phys. Today **63**, 27–32 (2010)

Y. Aharonov, L. Vaidman, Properties of a quantum system during the time interval between two measurements. Phys. Rev. A **41**(1), 11–20 (1990)

Y. Aharonov, L. Vaidman, Complete description of a quantum system at a given time. J. Phys. A: Math. Gen. **24**, 2315–2328 (1991)

Y. Aharonov, L. Vaidman, The two-state vector formalism: an updated review. Lect. Notes. Phys. **734**, 399–447 (2008)

V.I. Arnold, *Mathematical Methods of Classical Mechanics, Graduate Texts in Mathematics*, 2nd edn. (Springer, 1989)

M.V. Berry, P. Shukla, Typical weak and superweak values. J. Phys. A: Math. Theor. **43**, 354024 (2010)

S.M. Binder, Two-point characteristic function for the Kepler-Coulomb problem. J. Math. Phys. **16**(10), 2000–2004 (1975)

P. Boggiatto, E. Cordero, K. Gröchenig, Generalized anti-wick operators with symbols in distributional Sobolev spaces. Integr. Eqn. Oper. Theory **48**(4), 427–442 (2004)

M.A. de Gosson, *Born–Jordan Quantization*, Fundamental Theories of Physics 182, DOI 10.1007/978-3-319-27902-2

P. Boggiatto, G. De Donno, A. Oliaro, Time-frequency representations of Wigner type and pseudo-differential operators. Trans. Am. Math. Soc. **362**(9), 4955–4981 (2010)
P. Boggiatto, B.K. Cuong, G. De Donno, A. Oliaro, Weighted integrals of Wigner representations. J. Pseudo-Differ. Oper. Appl. (2010)
P. Boggiatto, G. Donno, A. Oliaro, Hudson's theorem for τ-Wigner transforms. Bull. London Math. Soc **45**(6), 1131–1147 (2013)
M. Born, P. Jordan, Zur Quantenmechanik. Zeits. Physik **34**, 858–888 (1925)
M. Born, W. Heisenberg, P. Jordan, Zur Quantenmechanik II. Z. Physik **35**, 557–615 (1925); English translation in: M. Jammer, *The Conceptual Development of Quantum Mechanics* (McGraw-Hill, New York, 1966); 2nd edn. (American Institute of Physics, New York, 1989)
V.C. Buslaev, Quantization and the W.K.B. method. Trudy Mat. Inst. Steklov **110**, 5–28 (1978) (in Russian)
M.G. Calkin, R. Weinstock, *Lagrangian and Hamiltonian Mechanics* (World Scientific, Singapore, 1996)
L. Castellani, Quantization rules and Dirac's correspondence Il Nuovo. Cimento **48A**(3), 359–368 (1978)
J. Chazarain, A. Piriou, *Introduction à la théorie des équations aux dérivées partielles linéaires* (Gauthier–Villars, Paris, 1981). English translation: *Introduction to the theory of linear partial differential equations. Studies in Mathematics and its Applications*, vol. 14 (North–Holland, 1982)
A.J. Chorin, T.J. Hughes, M.F. McCracken, J.E. Marsden, Product formulas and numerical algorithms. Commun. Pure Appl. Math. **31**(2), 205–256 (1978)
L. Cohen, Generalized phase-space distribution functions. J. Math. Phys. **7**, 781–786 (1966)
L. Cohen, Hamiltonian operators via Feynman path integrals. J. Math. Phys. **11**(11), 3296–3297 (1970)
L. Cohen, Correspondence rules and path integrals. J. Math. Phys. **17**(4), 597–598 (1976)
L. Cohen, *The Weyl Operator and Its Generalization* (Springer Science & Business Media, 2012)
L. Cohen, Can quantum mechanics be formulated as a classical probability theory? Philos. Sci. **33**(4), 317–322 (1966)
J.V. Corbett, The Pauli problem, state reconstruction and quantum-real numbers. Rep. Math. Phys. **57**(1), 53–68 (2006)
H.O. Cordes, On compactness of commutators of multiplications and convolutions, and boundedness of pseudodifferential symbols. J. Funct. Anal. **18**, 115–131 (1975)
E. Cordero, M. de Gosson, F. Nicola, On the invertibility of Born–Jordan quantization. Preprint 2015, (2015). arXiv:1507.00144 [math.FA]
P. Crehan, The parametrisation of quantisation rules equivalent to operator orderings, and the effect of different rules on the physical spectrum. J. Phys. A: Math. Gen. 811–822 (1989)
J.P. Dahl, M. Springborg, Wigner's phase space function and atomic structure: I. The hydrogen atom ground state. Mol. Phys. **47**(5), 1001–1019 (1982)

J.P. Dahl, W.P. Schleich, Concepts of radial and angular kinetic energies. Phys. Rev. A **65**(2), 022109 (2002)
T.G. Dewey, Numerical mathematics of Feynman path integrals and the operator ordering problem. Phys. Rev. A **42**(1), 32–37 (1990)
H.B. Domingo, E.A. Galapon, Generalized Weyl transform for operator ordering: polynomial functions in phase space. J. Math. Phys. **56**, 022104 (2015)
C. Eckart, Operator calculus and the solution of the equation of quantum dynamics. Phys. Rev. **28**, 711–726 (1926)
W.A. Fedak, J.J. Prentis, The 1925 Born and Jordan paper "On quantum mechanics". Am. J. Phys. **77**(2), 128–139 (2009)
H.G. Feichtinger, Un espace de Banach de distributions tempérées sur les groupes localement compact abéliens. C.R. Acad. Sci. Paris, Série A–B **290**(17), A791–A794 (1980)
H.G. Feichtinger, On a new Segal algebra. Monatsh. Math. **92**(4), 269–289 (1981)
H.G. Feichtinger, *Banach Spaces of Distributions of Wiener's Type and Interpolation*, Functional Analysis and Approximation (Oberwohlfach, 1980). Internat. Ser. Numer. Math. 60 (Birkhäuser, Basel, 1981), pp. 153–165
H.G. Feichtinger, Modulation spaces: looking back and ahead. Sampling Theory Signal Image Process. **5**(2), 109–140 (2006)
G.B. Folland, *Harmonic Analysis in Phase Space. Annals of Mathematics Studies* (Princeton University Press, Princeton, 1989)
C. Garrod, Hamiltonian path-integral methods. Rev. Mod. Phys. **38**(3), 483–494 (1966)
I.M. Gelfand, D.B. Fairlie, The algebra of Weyl symmetrised polynomials and its quantum extension. Commun. Math. Phys. **136**(3), 487–499 (1991)
H. Goldstein, *Classical Mechanics* (Addison–Wesley, 1950); 2nd edn. (1980); 3rd edn. (2002)
M. de Gosson, Maslov indices on the metaplectic group Mp(n). Ann. Inst. Fourier **40**(3), 537–555 (1990)
M. de Gosson, *Maslov Classes, Metaplectic Representation and Lagrangian Quantization*, Mathematical Research, vol. 95 (Wiley VCH, 1997)
M. de Gosson, *The Principles of Newtonian and Quantum Mechanics: The Need for Planck's Constant, h* (With a foreword by B. Hiley) (Imperial College Press, World Scientific, 2001)
M. de Gosson, The Weyl representation of metaplectic operators. Lett. Math. Phys. **72**, 129–142 (2005)
M. de Gosson, S. de Gosson. An extension of the Conley–Zehnder index, a product formula and an application to the Weyl representation of metaplectic operators. J. Math. Phys. **47**(12), 123506, 15pp. (2006)
M. de Gosson, *Symplectic Geometry and Quantum Mechanics*. Operator Theory: Advances and Applications. Advances in Partial Differential Equations, vol. 166 (Birkhäuser, Basel, 2006)
M. de Gosson, Metaplectic representation, Conley-Zehnder index, and Weyl calculus on phase space. Rev. Math. Phys. **19**(8), 1149–1188 (2007)

M. de Gosson, On the usefulness of an index due to Leray for studying the intersections of Lagrangian and symplectic paths. J. Math. Pures Appl. **91**, 598–613 (2009)
M. de Gosson, F. Luef, On the usefulness of modulation spaces in deformation quantization. J. Phys A: Math. Theor. **42**(1), 315205 (2009)
M. de Gosson, *Symplectic Methods in Harmonic Analysis and in Mathematical Physics* (Birkhäuser, 2011)
M. de Gosson, S. de Gosson, The reconstruction problem and weak quantum values. J. Phys. A: Math. Theor. **45**(11), 115305 (2012)
M. de Gosson, S. de Gosson, Weak values of a quantum observable and the cross-Wigner distribution. Phys. Lett. A **376**(4), 293–296 (2012)
M. de Gosson, L.D. Abreu, Weak values and Born-Jordan quantization. Quantum Theor: Reconsiderations Found. **6**(1508), 156–161 (2012)
M. de Gosson, Symplectic covariance properties for Shubin and Born-Jordan pseudo-differential operators. Trans. Am. Math. Soc. **365**(6), 3287–3307 (2013)
M. de Gosson, Born-Jordan quantization and the equivalence of the Schrödinger and Heisenberg pictures. Found. Phys. **44**(10), 1096–1106 (2014)
M. de Gosson. From Weyl to Born–Jordan quantization: The Schrödinger Representation Revisited. Phys. Rep. 2015 [in print]
M.J. Gotay, H.B. Grundling, G.M. Tuynman, Obstruction results in quantization theory. J. Nonlinear Sci. **6**, 469–498 (1996)
M.J. Gotay, On the Groenewold-Van Hove problem for R^{2n}. J. Math. Phys. **40**(4), 2107–2116 (1999)
A. Grossmann, Parity operators and quantization of δ-functions. Commun. Math. Phys. **48**, 191–193 (1976)
K. Gröchenig, Time-frequency analysis on Sjöstrand's class. Rev. Mat. Iberoamericana. **22**(2), 703–724 (2006)
K. Gröchenig, *Foundations of Time-Frequency Analysis* (Birkhäuser, Boston, 2000)
W. Heisenberg, Über quantentheoretische Umdeutung kinematischer und mechanischer Beziehungen. Z. Physik **33**, 879–893 (1925)
B.J. Hiley, Weak values: approach through the clifford and moyal algebras. J. Phys.: Conf. **361**(1) (IOP Publishing, 2012)
L. van Hove, Sur certaines représentations unitaires d'un groupe fini de transformations. Mém. Acad. Roy. Belg. Classe des Sci. **26**(6) (1952)
L. Hörmander, On the division of distributions. Ark. Mat. **3**, 555–5568 (1958)
L. Hörmander, *The Analysis of Linear Partial Differential Operators*, vol. I (Springer, 1981)
L. Hörmander, *The Analysis of Linear Partial Differential Operators* I. Grundl. Math. Wissenschaft. 256 (Springer, 1983)
L. Hörmander, *The Analysis of Linear Partial Differential Operators III* (Springer, Berlin, 1985)
T. Kato, On the Trotter-Lie product formula. Proc. Jpn Acad. **50**, 694–698 (1976)
Y. Katznelson, *An introduction to Harmonic Analysis* (Dover, 1976)
S.K. Kauffmann, Getting path integrals physically and technically right (2010). arXiv:0910.2490 [physics.gen-ph]

S.K. Kauffmann, Unambiguous quantization from the maximum classical correspondence that is self-consistent: the slightly stronger canonical commutation rule dirac missed. Found. Phys. **41**(5), 805–819 (2011)

E.H. Kerner, W.G. Sutcliffe, Unique Hamiltonian operators via Feynman path integrals. J. Math. Phys. **11**(2), 391–393 (1970)

J. Leray, *Lagrangian Analysis and Quantum Mechanics*. A Mathematical Structure Related to Asymptotic Expansions and the Maslov Index (the MIT Press, Cambridge, Mass., 1981); translated from Analyse Lagrangienne RCP 25 (Strasbourg Collège de France, 1976–1977)

R.G. Littlejohn, The semiclassical evolution of wave packets. Phys. Rep. **138**(4–5), 193–291 (1986)

F. Luef, Z. Rahbani, On pseudo-differential operators with symbols in generalized Shubin classes and an application to Landau-Weyl operators. Banach J. Math. Anal. **5**(2), 59–72 (2011)

J.S. Lundeen, B. Sutherland, A. Patel, C. Stewart, C. Bamber, Direct measurement of the quantum wavefunction. Nature **474**(7350), 188–191 (2011)

N. Makri, W.H. Miller, Correct short time propagator for Feynman path integration by power series expansion in Δt. Chem. Phys. Lett. **151**, 1–8 (1988)

N. Makri, Feynman path integration in quantum dynamics. Comput. Phys. Commun. **63**(1), 389–414 (1991)

V.P. Maslov, M.V. Fedoriuk, *Semi-classical Approximation in Quantum Mechanics*, vol. 7. (Springer Science & Business Media, 2001) (D. Reidel, Boston, 1981)

V.P. Maslov, V.C. Bouslaev, V.I. Arnol'd, *Théorie des perturbations et méthodes asymptotiques* (Dunod, 1972)

N.H. McCoy, On the function in quantum mechanics which corresponds to a given function in classical mechanics. Proc. Natl. Acad. Sci. U.S.A. **18**(11), 674–676 (1932)

F.G. Mehler, Über die Entwicklung Einer function von beliebig vielen Variabeln nach Laplaceschen Functionen höherer Ordnung. J. für Reine und Angew. Math. (in German) **66**, 161–176 (1866)

J. Mehra, H. Rechenberg, *The Historical Development of Quantum Theory*, vol. 1. The Quantum Theory of Planck, Einstein, Bohr, and Sommerfeld: Its Foundation and the Rise of Its difficulties (Springer-Verlag, Berlin, 1980)

C.L. Mehta, Phase-space formulation of the dynamics of canonical variables. J. Math. Phys. **5**(1), 677–686 (1963)

S.P. Misra, T.S. Shankara, Semiclassical and quantum description. J. Math. Phys. **9**(2), 299–304 (1968)

J. von Neumann, *Mathematical Foundations of Quantum Mechanics* (Princeton University Press, Princeton, 1955)

F. Nicola, L. Rodino, *Global Pseudo-Differential Calculus on Euclidean Spaces*. Pseudo-Differential Operators (Birkhäuser, 2010)

J. Niederle, J. Tolar, Quantization as mapping and as deformation. Czech. J. Phys. **B29**, 1358–1368 (1979)

D. Park, *Introduction to Quantum Theory*. International Series in Pure and applied Physics (McGraw-Hill, 1992)

W. Pauli, *Die allgemeinen Prinzipen der Wellenmechanik*. Handbuch der Physik, vol. 5 (Springer, Berlin, 1958)
L. Pauling, *General Chemistry*, 3rd edn. (W.H. Freeman & Co., 1970), p. 125
A. Peres, Unperformed experiments have no results. Am. J. Phys. **46**(7), 745–747 (1978)
M. Przanowski, J. Tosiek, Weyl–Wigner–Moyal formalism. I. Operator ordering. Acta. Phys. Pol. B **26**, 1703–1716 (1995)
H. Reiter, Metaplectic groups and Segal algebras. Lecture Notes Math. **1382** (1989)
N.W.M. Ritchie, J.G. Story, R.G. Hulet, Realization of a measurement of a "Weak Value". Phys. Rev. Lett. **66**(9), 1107–1110 (1991)
A. Royer, Wigner functions as the expectation value of a parity operator. Phys. Rev. A **15**, 449–450 (1977)
E. Schrödinger, Das Verhältnis der Heisenberg-Born-Joordanschen Quantenmechanik zu der meinen. Annalen der Physik **79**, 734–756 (1926)
E. Scholz, *Weyl Entering the 'new' Quantum Mechanics Discourse*, ed. by C. Joas, C. Lehner, J. Renn. HQ-1: Conference on the History of Quantum Physics (Berlin, July 2–6, 2007). Preprint MPI History of Science, 350 vol. II (Berlin, 2007)
E. Segal, Foundations of the theory of dynamical systems of infinitely many degrees of freedom. Mat. fys. Medd. Danske Vid. Selsk **31**(12), 1–39 (1959)
D. Shale, Linear symmetries of free boson fields. Trans. Am. Math. Soc. **103**, 149–167 (1962)
J.R. Shewell, On the formation of quantum-mechanical operators. Am. J. Phys. **27**, 16–21 (1959)
A. Shimony, *The Search for a Naturalistic World View*, vol. 1 (Cambridge University Press, 1993)
M.A. Shubin, *Pseudodifferential Operators and Spectral Theory* (Springer, 1987) (original Russian edition in Nauka, Moskva 1978)
J. Sjöstrand, An algebra of pseudo-differential operators. Math. Res. Lett. **1**(2), 185–192 (1994)
J. Sjöstrand, *Wiener Type Algebras of Pseudo-Differential Operators*, Séminaire sur les Équations aux Dérivées Partielles, École Polytech., Palaiseau, Exp. No. IV, 21 (1994–1995)
M.H. Stone, On one-parameter unitary groups in Hilbert space. Ann. Math. Second Ser. **33**(3), 643–648 (1932)
J.L. Synge, C. Truesdell, *Handbuch der Physik: Prinzipien der Klassischen Mechanik und Feldtheorie*, vol. 3 (Springer, 1960)
J. Toft, Multiplication properties in pseudo-differential calculus with small regularity on the symbols. J. Pseudo-Differ. Oper. Appl. **1**, 101–138 (2010)
J. Toft, Continuity properties for modulation spaces, with applications to pseudo-differential calculus. I.J. Funct. Anal. **207**(2), 399–429 (2004)
K. Vo-Khac, *Distributions, Analyse de Fourier, Opérateurs aux Dérivées Partielles, Tome 2* (Vuibert France, 1972)
N. Wallach, *Lie Groups: History, Frontiers and Applications*, vol. 5, Symplectic Geometry and Fourier Analysis (Math. Sci. Press, Brookline, 1977)

G.N. Watson, Notes on generating functions of polynomials: (2) Hermite polynomials. J. Lond. Math. Soc. **8**, 194–199 (1933)

A. Weil, Sur certains groupes d'opérateurs unitaires. Acta Math. **111**, 143–211 (1964)

H. Weyl, Quantenmechanik und Gruppentheorie. Zeitschrift für Physik **46** (1927)

H. Weyl, *The Theory of Groups and Quantum Mechanics*, translated from the 2nd German edition by H.P. Robertson (Dutten, New York, 1931)

M.W. Wong, *Weyl Transforms* (Springer, 1998)

W.H. Zurek, Sub-Planck structure in phase space and its relevance for quantum decoherence. Nature **412**(6848), 712–717 (2001)

Index

M.A. de Gosson, *Born–Jordan Quantization*, Fundamental Theories of Physics 182, DOI 10.1007/978-3-319-27902-2

Printed in the United States
By Bookmasters